Behavioural Problems in Rabbits

Behavioural Problems in Rabbits

A Clinical Approach

Guen Bradbury, MA, VetMB, MRCVS

5m Publishing

First published 2018

Copyright © Guen Bradbury 2018

Published by
5M Publishing Ltd,
Benchmark House,
8 Smithy Wood Drive,
Sheffield, S35 1QN, UK
Tel: +44 (0) 1234 81 81 80
www.5mpublishing.com

A Catalogue record for this book is available from the British Library

ISBN 9781789180121

Book layout by Servis Filmsetting Ltd, Stockport, Cheshire
Printed by Replika Press Pvt. Ltd., India
Photos and illustrations by the author unless otherwise indicated

Dedication

To Greg, my co-author, rabbit photographer, and husband.

Contents

Preface

When I was 11, my family found a stray rabbit that had been abandoned next to a dual carriageway. My siblings and I captured the rabbit, a Siamese-point Netherland Dwarf who came to be called Bun. As his primary carer, I read everything I could find about keeping rabbits. We built him a hutch, fed him on a varied diet that included muesli mix, and I carried him to a large run in the garden every day so he could graze. I could handle him without problems, I loved him very much, and he lived for ten years with my family.

After his death, I inherited another rabbit, Pewter. I tried to keep him in the same way that I had kept Bun, but he started to bite and scratch anyone who approached his hutch, froze when he was lifted, and tried to avoid human interaction as much as possible. He was clearly unhappy, but I didn't know how to improve his welfare.

Not long after that, I started at veterinary school. Although we were trained how to modify behaviour in many companion animal species, little mention was made of rabbit behaviour. So, in addition to our lectures, I combed the academic literature for advice — reading everything from studies investigating rabbit environmental preferences, to the many ways that diet affects rabbit behaviour. I then needed to put this knowledge into practice.

In my first veterinary job, I ran behaviour clinics for a variety of species, but I still rarely saw rabbits. I found that unwanted behaviours in these animals are rarely considered a problem by the owner: the signs of fear and distress are not immediately obvious, and rabbits almost never injure owners badly enough for them to seek help on this account.

I needed a way to reach rabbit owners that was engaging and allowed owners to contact me directly. So, I set up a YouTube channel and posted videos of my rabbits doing trained trick behaviours. Underneath each video, I described how the tricks were trained, and I encouraged rabbit owners to post comments or send me private messages. I made a point of responding to every contact. I started to get general comments, 'Can't wait to get my first rabbit' – this gave me the chance to start a conversation about how rabbits are happier kept in pairs. I then started getting very specific messages, 'My rabbit bites me when I give her food,' 'I need to move my rabbit 5000 miles, how do I prevent him getting stressed on the journey,' and 'How do I make my rabbit love me more?'

One thing led to another. I set up a website on rabbit behaviour and, before long, owners were contacting me with very specific questions. Some were answered via email, some over the phone, and some through video calls. All of these interactions gave me the chance to apply the knowledge and techniques that I learned. They also taught me something important: the majority of behavioural problems in rabbits occur because of the owner's inappropriate interactions with, or expectations of, their rabbits.

To get some new ideas into veterinary discussion, I co-authored various papers, articles and letters on the subject of rabbit behaviour in veterinary journals. While this was great for engaging with the profession, owners don't often have access to this information. And it is for exactly that reason that I wrote this book – to pull all of the best evidence together and act as a best-practice instruction manual for interested owners, and a reference guide for vets who want to help their clients and their pets.

While I thought I was giving Bun the best life possible, I made a lot of assumptions. If I had the chance to care for him again, I would do things differently. Hindsight can be a terrible thing. I don't want anyone to make the same mistakes as me. Not all rabbits that are kept in suboptimal ways will show behavioural problems, but all rabbits have similar wants and needs. Improving their behavioural health will improve their welfare, and often improving their welfare will improve their behaviour. This book is for readers who want to understand rabbit behaviour problems, and want simple, evidence-based advice to improve them.

With Pewter, I decided to make three changes: I bonded him to a companion, Babbitt; I trained him to perform some simple cued behaviour, so I never needed to pick him up; and I started to 'ask' him whether he wanted me to interact with him. He's now nine years old, has outlived Babbitt, and is happily bonded to a second companion, Terne. He hasn't bitten anyone for five years. He comes up and nudges me to stroke his head. He sits next to me and licks me. And he is patient and tolerant of my young daughter, while she, too, learns her rabbit manners.

I wish you all the same success when dealing with your own rabbit behaviour problems.

Key points

To successfully solve behavioural problems in rabbits, using a clinical approach:

1. Repair the husbandry

Solving behavioural problems is much easier when the rabbits' husbandry is ideal. This means:

🐰 **Companionship**

Rabbits must not be kept without a rabbit companion.

🐰 **Grass**

A suitable rabbit diet is mostly, or entirely, grass.

🐰 **Outdoors**

Rabbits do better with access to the outdoors.

🐰 **Grounded**

Being picked up scares rabbits and worsens behaviour.

🐰 **Choice**

It is vital that rabbits have choice over their interactions with humans.

2. Get in training

When rabbits can cope with their environment, they will behave more normally. Training can help them cope.

3. Set appropriate expectations

Rabbits are a relatively recently domesticated prey species. Their motivations, interactions and normal behaviours are very different from cats and dogs. Owners must be aware of this.

4. Take small steps

Try to set an achievable behaviour modification plan. If many steps are required, build up gradually and highlight results along the way. Remember, the least successful behavioural plan is the one that is not followed...

Introduction

You have picked up this book, so you obviously have an interest in rabbit behaviour problems. Maybe you're a veterinary surgeon or in a position of responsibility and people come to you for expert advice? Maybe you are encountering unwanted behaviour in your own rabbit? Maybe you've studied unwanted behaviours in different species and you're wondering how these differ in rabbits? This book will help you to understand rabbit behaviour problems, and give simple, evidence-based advice to improve them.

Let's imagine that a friend of yours has a rabbit that she loves dearly, and asks you for some advice. Her rabbit is often affectionate towards her, but occasionally bites and scratches when she tries to get him out of his hutch. She doesn't understand why the rabbit does this. Her internet searches suggest that she puts on gloves when she handles him, that she needs to show him that biting doesn't work, that she should handle him more so he becomes accustomed to it. Her vet recommends that she get the rabbit neutered. She's asked you whether you have any suggestions.

You have some ideas – you can see that there are a few aspects of the way that this rabbit is kept that are not perfect. But how do you know which aspects are likely to be contributing the most? Which aspects should you suggest that your friend changes first? And will those husbandry changes alone resolve the problem?

This book contains practical, useful advice on resolving unwanted behaviours in rabbits: whether that's for your friend, your client or even yourself. You can read it from cover to cover to build a structured understanding of rabbit welfare, motivations and the general principles of rabbit learning, or you can dip in and out of it to find advice on managing specific unwanted behaviours.

I have set up the following chapters to mirror a basic behavioural consultation:

1. **Introduction**
    This section gives an overview of how humans interact with rabbits: how the pet rabbit was domesticated, what the societal expectations of pet rabbits are, and how different ownership styles affect an owner's relationship with their pet. This will help you to manage the owner's behaviour as well as the rabbit's, setting up the consultation to lead to an outcome that the owner can realistically achieve.

2. **What is the problem behaviour?**
    This section covers the sort of information that you need to get from the owner to address a problem behaviour, and the various tools you can use to help you to get the comprehensive history that you need.

3. **Why is the problem behaviour happening?**
    This section aims to help you to get to a diagnosis. It is structured around a framework for assessing welfare in rabbits, and considers the diet, environment, health, and ability to assess normal behaviour. These factors inform how the rabbit

interprets its situation. Understanding what rabbits need will help you to make sense of what is missing in an individual rabbit's husbandry.

4. **How can I change behaviour?**
    This section introduces the general principles of management of an unwanted behaviour. It provides advice on how to help owners to read their rabbit's behaviour more carefully, so they can adjust their interactions appropriately. It explains how to change a rabbit's situation with minimal stress. Additionally, it describes methods of training rabbits, gives tips and suggests behaviours that should be trained to help the rabbit to cope with its environment.

5. **How can I resolve a specific behavioural problem?**
    This section gives advice on specific behavioural problems, categorised as those that occur between the rabbit and its owner; between the rabbit and its companion rabbit; between the rabbit and its environment; and those that are self-directed.

6. **Conclusion**
    This section briefly recaps and reminds you of the key points.

Finally, there is a useful appendix of advice sheets for owners on different aspects of behavioural or environmental modification that have been covered within the text. These can be photocopied and handed out to owners where appropriate.

Returning to our hypothetical situation, let's imagine that you've now read this book. You feel much more confident about the advice that you want to give your friend. You can ask her about what she feels a good outcome would be. You can discuss how rabbits feel about being handled. You can talk about environmental modifications and you can help her understand what her rabbit is trying to communicate. You

can give specific advice to help her to resolve this problem. When she encounters hurdles in the process, you can discuss them and suggest how these barriers can be overcome. As she finds her relationship with her rabbit beginning to improve, she's likely to become more motivated to improve its welfare. You can then advocate other improvements in the rabbit's welfare, such as the introduction of a rabbit companion. You have improved the welfare of both the rabbit, and of your friend as well.

It is hard for humans to communicate with prey species like rabbits. Evolutionarily, humans are predators, and so we have not evolved to understand the subtleties of communication of prey species. This means that understanding their motivations and drives requires more work and is less intuitive. However, the skills that we can learn (patience, building trust, recognising subtle signals, developing a balanced relationship) are very transferable. Understanding rabbit behaviour can give us new insights into how we interact with other animals and even how we interact with our fellow humans.

Good luck! I hope you enjoy the book. If you have any questions, please don't hesitate to get in touch: my contact details are at the back.

Changing expectations

In this section, we'll explore why there is frequently a mismatch between how rabbits want to behave around humans, and how humans want to behave around rabbits.

Rabbits have been popular children's pets for many years, but the demographic of rabbit owners, and their expectation of their rabbits, is changing. The 'in a hutch, in a shed' model of husbandry, though regrettably still seen, no longer represents the norm. These changing expectations have improved rabbit welfare in a number of ways, but can place other stresses on these animals. Many owners now expect their rabbits to live in the house without destroying it, to enjoy frequent, unsolicited cuddles without complaint and to display strong, affectionate relationships with humans.

The 2016 PDSA Animal Wellbeing (PAW) survey reported that 3% of homes in the UK have a rabbit, with an estimated population of 1.5 million rabbits. About 25% of rabbits were purchased because the children wanted a pet.

Considering an animal's natural history can provide some context to its behaviour. Unlike wild dogs and cats, the wild rabbit is a herbivorous prey species. This species forms more than 50 per cent of the diet of more than 30 different predators. This alarming statistic explains a lot of the rabbit's behaviour in captivity. When thinking about any rabbit behaviour, it is worth remembering that their ingrained primary motivation is to avoid being eaten!

Rabbits can make very good pets. They require relatively little human interaction if kept in same-species pairs or groups, they are more eco-friendly than cats and dogs (herbivores rather than carnivores), they can be easily provided with a diet that is close to their natural diet, they do not require the same degree of interaction and care when the owner goes away (less dependent on human companionship), they are quiet (fewer problems with neighbours) and they can be friendly and affectionate if well managed.

In the 2016 PAW report, rabbit owners were significantly less likely than dog or cat owners to say that owning their pet made them happy.

Although many owners expect that the rabbits that they acquire will be interactive and low-maintenance pets, it seems that the expectation is frequently not met. The experience of owning rabbits may be unrewarding and frustrating for some owners. Owners may describe their rabbit as 'boring' (reflecting insufficient opportunity to express normal behaviours), 'timid' (reflecting husbandry and interactions that fail to meet the rabbit's emotional needs) or 'grumpy' (reflecting a lack of awareness of rabbit behaviour resulting in expression of aggressive behaviour). Owners who have problems with their rabbit's behaviour may find it very hard to get appropriate advice. Vets aren't yet taught about rabbit behaviour at university, there are few behaviourists who work specifically with rabbits, and internet advice is often conflicting.

When asked specifically about behavioural problems in their rabbit, 50% of rabbit owners in the 2016 PAW report said they would go to a veterinary professional for help, 38% would rely on internet advice, and 14% of owners wouldn't seek help.

So, it's clear that there is a general misunderstanding about what rabbits need from their environment and their interactions, and there is very little training available to interested individuals to improve their understanding. Remedying that by digging into the ecology, evolution and history of the rabbit–human relationship will show us the context within which modern domestic rabbit behaviours occur. We'll start by thinking about how humans, as a species, have ended up keeping rabbits as pets.

Domestication

This section describes how rabbits were domesticated, and why this affects their behaviour in captivity.

Domestication refers to the long-term process in which one species controls various aspects of care and reproduction of another species, which results in a change in the second species that makes it more valuable to the first species. This process generally makes the domesticated species more able to cope with life in captivity and less well adapted to life in the wild.

It is helpful to compare rabbit domestication to that of other pet species: dogs were domesticated (i.e. began to be appreciably different from the wild grey wolves) between 40,000 and 100,000 years ago; cats were domesticated about 10,000 years ago; and rabbits were domesticated about 1,400 years ago. Dogs have been domesticated for much longer than rabbits, which is part of the reason that humans find it inherently easier to read dog behaviour.

However, this ability to understand a species' behaviour does not just depend on how long the animal species has been domesticated. Route and reason of domestication also makes a difference. The ancestors of domestic dogs probably scavenged around human habitation. The animals with a lower flight threshold (i.e. those that were tamer) could get closer to humans and therefore got more food. These scavenging dogs probably started to alert humans when danger threatened, so humans realised that a close relationships with dogs was valuable. This relationship benefited both humans and dogs. When humans encouraged dogs to live close by, the tamer animals would have an advantage over the more fearful ones. Selection for specific traits then became more common, meaning dogs could be bred for hunting, guarding, herding, hauling, killing vermin and, more recently, appearance.

These differing selection pressures resulted in the many dog breeds that exist today.

Rabbits, on the other hand, were domesticated in AD 600, initially as a low-maintenance food source, and subsequently for their fur. As a result, the selected traits were body size and coat characteristics (Figure 1.1). Unlike dogs, rabbits have never been deliberately selected for behavioural traits.

So, while dogs were domesticated a long time ago, and were mostly selected on behavioural characteristics, rabbits were domesticated relatively recently, and were mostly selected on physical characteristics. As a result, their instinctive responses and range of behaviours are very different. The relationships that dogs and rabbits can form with humans are very different and humans should interact with them accordingly.

Looking in more detail, dogs appear to have a relationship with humans analogous to the parent–child relationship. A famous study (Belyaev, 1979) involved the domestication of silver foxes. The scientists bred litters of silver foxes, and then bred the tamest animals together (those that were most willing to approach the scientists). Within 20 generations, the foxes were more like dogs. Not only did they behave more like dogs (wagging the tail, licking their caretakers, showing reduced fear responses), they also looked more like dogs (floppy ears, short or curly tails, changes in the shape of the skull). These behavioural and appearance changes are similar to those of juveniles, but are retained into adulthood. It seems that dogs may be 'neotenous' wolves. They retain puppy-like physical and behavioural features into adulthood, and they show more attachment to human carers than even the tamest, hand-reared, wild-type wolves.

The parent–child similarity in the dog–human relationship is unique among our relationships with animals. It is clear that the

Figure 1.1: Many rabbits, such as this Flemish Giant (which weighs about 8 kg), were selected for size and meat quality.

process of domestication of rabbits, which has been short and based on traits other than tameness, has not selected for a similar relationship. A dog is emotionally dependent on its human caretakers. A rabbit may be physically dependent on its human caretakers, but it does not have the same intrinsic requirement for humans to provide emotional stability.

This is not to say that humans and rabbits cannot form strong, mutually beneficial relationships, but owners who base their expectations of pet rabbits on their experiences with dogs are likely to be disappointed. Rabbits are less likely to 'forgive' errors, because they do not depend on humans to the same degree that dogs do. This means that any training must be better executed, and that trust and predictability are vital to ensure that the human–animal relationship is sustainable.

This relationship between animal and owner affects the approach to behavioural problems. As a contrasting example to the dog, consider how behaviours are modified in the cat. When changing a behaviour in a cat, modifying the environment is usually more important than altering the owner's interactions or offering play as a reward. Similarly, in rabbits, modifying the environment and companionship status is very important.

An animal's behaviour can be categorised in a number of different ways: normal versus abnormal; wanted versus unwanted. The majority of 'problem behaviours' in rabbits are unwanted, normal behaviours. Since humans don't have an instinctive understanding of rabbit behaviour, we often have unrealistic expectations of rabbits. Managing owners' expectations is key to reducing the number of normal behaviours that are unexpected, and therefore unwanted. A mismatch between an owner's expectation and a rabbit's behaviour underlies most of the 'problem behaviours' that owners report. There are also some 'behavioural disorders', abnormal behaviours, such as stereotypies.

Throughout this book, I use the phrase 'problem behaviours' to refer to any behaviours, normal or abnormal, that are causing the owner a problem. The phrase 'unwanted behaviours' is used for those behaviours that are normal, but pose a problem to the owner, and 'abnormal behaviours' for those behaviours that are not part of the wild rabbit's normal behavioural repertoire.

Also, when I compare rabbits with another pet species, I will compare them with dogs. I will, for example, contrast and compare the relationships that rabbits form, and the husbandry that they receive, to that of dogs. This choice is both because dogs are very familiar and most readers will instantly be able to picture dog behaviour, and also because the species are very different.

Considerations for behaviour modification in rabbits

Rabbit owners are less likely to seek help to resolve 'problem behaviours' in their pet, and they may be less likely to act on behavioural modification advice than the owners of dogs.

There are various reasons for this.

1. The human–animal bond (the mutually beneficial, dynamic bond between an individual person and animal) is usually weaker between rabbits and their owners than between dogs and theirs.
2. Culturally, rabbits are seen as cheap, low-maintenance pets.
3. Unwanted behaviours in rabbits usually cause less direct inconvenience or injury to the owner than unwanted behaviours in dogs.
4. Finally, there is little education available

to owners about how to manage unwanted behaviours in rabbits.

The 2016 PAW report found that 45% of veterinary surgeons were concerned that a key problem for rabbits was a complete lack of care (i.e. rabbits being forgotten about). A total of 59% of rabbit owners would like to spend more time with their pet than they currently do – significantly more than dog (50%) and cat (46%) owners. Reasons for not being able to do this are cited as work hours (65%), family commitments (23%) and social activities (19%).

Let's examine these reasons in some detail:

Weaker human–animal bond

The strength of the human–animal bond between a pet and its owner can be measured by questionnaire for the human, and motivation and separation studies for both parties. A variety of sources (e.g. Bonas et al., 2000; Cromer and Barlow, 2013) have shown that the bond is usually weaker between rabbit owners and their rabbits than between dog owners and their dogs. This reduces the likelihood of owners seeking help regarding their rabbits: a strong human–animal bond is a major motivator to improve their rabbits' welfare. Anecdotally, rabbit owners are less likely than dog owners to request to be present at the euthanasia of their pet and less likely to request that the ashes of their rabbit are returned after cremation. Strengthening the human–animal bond can improve a rabbit's welfare: increasing owner attachment correlates with regular veterinary check-ups and more knowledge of the rabbit's needs.

Cultural perception that rabbits are cheap

Partly because they were initially kept for meat and partly because they breed quickly, rabbits are cheap to buy in the UK and are culturally seen as disposable. People rarely spend a lot of money to maintain something that was cheap to purchase. Hence, behaviour modification in rabbits is frequently constrained by finances. In addition, few rabbits are insured, and even if they are, few insurance companies cover behavioural medicine.

Rabbit bites rarely cause severe injury

As rabbits are typically smaller than dogs and cats and often have less physical contact with the owners, unwanted behaviours are less likely to directly inconvenience the owner. Even in the case of aggressive behaviours, the injury risk from a rabbit is much lower than that from a cat or a dog, so the risk may not be seen as sufficient reason to change the husbandry or interactions. If behaviours indicative of stress do not directly affect the owner, the owner will be unlikely to see these as a problem for the rabbits.

Low awareness of treatment options

Owners may also not realise that there are options to resolve behaviour problems in rabbits. Vets are often reluctant to advise on unwanted behaviours. Veterinary behavioural medicine is a new field (so many older vets have not been trained in the principles of behavioural modification), and very few vets receive any training specifically on behavioural problems in rabbits. Additionally, unlike in cats and dogs, behavioural modification of rabbits does not involve the use of pharmaceutical agents. There are no drugs licensed in rabbits to alter behaviour, so vets will not get continuing education through pharmaceutical reps.

Many sources of advice on rabbit behaviour have a very simplistic approach to finding or creating solutions. Owners are told to make an intervention without understanding why the rabbit is performing the behaviour in the context of its overall husbandry. The information is often conflicting, so even those owners who are highly motivated to help their rabbits may still struggle to find useful, humane advice. The information above will help you to understand some of the barriers between pet rabbits and appropriate behavioural modification.

How 'ownership styles' affect owner beliefs

In the 2017 PAW report, 43% of rabbit owners were 18–34 years old. In general, they were younger than dog and cat owners. About half of rabbit owners were male.

It's clear that rabbit owners, on average, are unlikely to seek advice on resolving problem behaviours. However, different categories of rabbit owners have differing ownership styles, and these ownership styles affect how the owner perceives their pet. When you give advice, you should understand what the owner expects from their pet and how they feel about their pet. This allows you to tailor your advice to increase the likelihood that the owner listens and acts upon it.

Parents manage their children with a variety of different 'parenting styles', categorised by both the degree of control that is exerted over the child's behaviour, and the degree of emotional attachment that the parent has to the child. This model has been applied to animal

	High control	Low control
High emotional attachment	Authoritative	Indulgent
Low emotional attachment	Authoritarian	Uninvolved

Diagram 1.1: Ownership styles can be categorised by the degree of emotional attachment to, and the degree of control over, their rabbit.

owners (German, 2015). A rabbit owner's 'ownership style' can be categorised in the same way – on the basis of their emotional attachment to their rabbits and the degree to which they try to control their behaviour (Diagram 1.1).

We can apply these parenting styles to rabbit owners, as 'ownership styles' (the general framework for how an owner engages with their rabbit) and 'ownership practices' (specific methods that the owner uses that conform with their ownership style).

Authoritative

Owners with a strong emotional attachment to their rabbit and high level of control over its behaviours are likely to use positive reinforcement techniques. They are more likely to see behaviours indicative of fear and distress and be able to modify their own behaviour appropriately. When provided with an explanation of why an unwanted behaviour is occurring, they are likely to be willing to adapt their behaviour to change their rabbit's behaviour.

Authoritarian

Owners with a low emotional attachment to their rabbit are likely to control the rabbit's behaviour through its environment, restricting its ability to perform unwanted behaviours rather than

modifying them if they occur. They are more likely to use aversive techniques if the rabbit performs an unwanted behaviour and are less likely to see the behaviour as a symptom of poor welfare in their rabbit. They are less likely to modify their own behaviour to change the rabbit's behaviour.

Indulgent

Owners with a strong emotional attachment to their rabbit but a low level of control over its behaviour are likely to provide the rabbit with many toys and treats. They are unlikely to use training techniques as they have very low expectations of their pet. However, they respect the needs of their rabbit, and may well be motivated to change their behaviour if it will improve their rabbit's welfare.

Uninvolved

Owners without either an emotional attachment to their rabbit, or a desire for control over its behaviour, may well neglect their rabbit, as they have little or no motivation to control it or understand it. These owners are therefore very unlikely to seek help regarding an unwanted behaviour in their pet. Many owners who acquire a rabbit as an 'entry-level' pet for their children may fall into this category. They have no expectation of reward for themselves from owning the rabbit.

In parenting, specific behaviours ('parenting practices') are easy to try, cease or change. However, if any specific 'parenting practice' conflicts with the general approach of that parent's 'parenting style', then the parent will be unlikely to maintain the 'parenting practice'. The same is true of 'ownership styles'.

Recognising the different 'ownership styles', and the implications of these styles on how the owners generally regard and treat their pets, helps you to target your advice and make your prognoses more realistic. It is likely to be very hard to motivate an uninvolved owner to start training their animal, but it may be possible to advise them on how they can increase their rabbit's quality of life by providing it with a companion, for example. Cerebrally, most owners want to do the 'right thing' for their animal, even if emotionally, they are not very attached.

Finally, in parenting, the authoritative style of parenting is associated with the most positive behavioural outcomes. Rabbit owners with a similar approach are likely to be most responsive to guidance and most motivated to resolve unwanted behaviours, recognising that these behaviours reflect suboptimal welfare in their rabbit.

How common are problem behaviours in rabbits?

Rabbit owners, on average, are less likely than dog owners to seek help for problem behaviours in their pet. Similarly, certain types of rabbit owner may be more or less likely to seek help. So, how commonly do problem behaviours in rabbits occur? And how does this affect their owners?

> The 2016 PDSA Animal Wellbeing (PAW) report stated that 43% of rabbit owners say that their rabbit displays at least one behaviour they'd like to change. The major behaviours that the owners disliked were thumping of the hind feet and repeatedly biting the bars of the run or hutch — both signs that indicate a problem with the rabbit's welfare.

The 2016 PAW report found that owners who said that they did want to change behaviours in their rabbit were more likely to:

- feel that owning a pet is harder work and more expensive than expected.
- feel that owning a pet makes them stressed.
- be feeding rabbit muesli as a main type of food or at least once a month.
- give treats because it makes their rabbit happy.
- have tried a weight loss diet with their rabbit.
- feel uninformed about a rabbit's ability to express normal behaviour.

The survey also found that owners who didn't want to change behaviours in their rabbits were more likely to:

- feel informed about their rabbit's behavioural needs to express normal behaviour.
- disagree that owning a pet makes them stressed or is harder work than they expected.
- spend more time researching buying a pet than any other household purchase.
- own older rabbits.

This last point is interesting, as a higher average age of rabbits in the sample implies that that group are living longer and are, therefore, more likely to be having their health and husbandry needs met appropriately.

It seems that owners who are more informed about rabbits are less likely to dislike an aspect of their rabbit's behaviour. This is to be expected — people, in many instances, react poorly to the unexpected. However, in that PAW report, the percentage of owners who would like to change an aspect of their rabbit's behaviour was lower than the percentage of rabbits kept in poor conditions (52 per cent of rabbits lived alone, 22 per cent lived in hutches that are too small, on average the rabbits spent 12 hours per day in the hutch, and 26 per cent did not have any interaction with humans on a daily basis — a significant problem if the rabbit is kept on its own). Acceptance of normal behaviour from rabbits is one thing, but acceptance of all behaviour is a failure to recognise abnormal behaviours. And that is a bad thing.

Empathising with the owner

The previous section described various barriers to achieving successful resolution of behavioural problems. Some of these are with the owner, not the rabbit. It's therefore very important to work with the owner, not against them. For behavioural consultations, even more so than medical consultations, a high degree of empathy and tact is crucial. This is for various reasons:

- The owner's husbandry and interactions with the animal are extremely likely to be the cause of the problem.
- Most owners want to do the best thing for their animal, and highlighting discrepancies between their intentions and the consequences may make them feel bad about themselves. Guilt and self-recrimination are bad for motivation and we want owners to be motivated to improve the situation.
- Any change is likely to cause disruption for the owner. Changes may
  - Alter their home and environment.
  - Alter the interactions that they have with the rabbit.
- The owner may feel irritated that they have a responsibility for an unrewarding pet that is causing them problems.

2

What is the problem behaviour?

The previous section detailed the essential background information for any behavioural consultation: how domestication has affected rabbits, why humans are intrinsically poor at understanding rabbit behaviour, and the barriers that owners face when trying to resolve unwanted behaviours. This section will cover the consultation process. By the end of the section, you'll understand the various tools you can use to understand and analyse the problem behaviour.

A good consultation starts with a detailed history of the animal and the problem, with the goal of understanding the underlying cause, not just the symptoms. The problem behaviours reported in the PAW report (previous section) are symptoms of other problems.

When dealing with a behaviour problem in a rabbit, it's important to consider several high-level questions:

1. What would the owner like to change about the rabbit's behaviour? If this is not resolved, regardless of any improvements in the animal's welfare, the owner will not see it as a success.
2. Why is the rabbit motivated to perform this behaviour? What does it want to achieve?
3. Why are these motivations present? What is it about the environment or the interactions of the rabbit that create these motivations?
4. How can the motivation of the rabbit be changed? How can the problem behaviour that the owner is encountering be resolved?

To consider these high-level questions, there are several tools that can be used. The first, and most important, is a complete history from the owner.

Taking a history

Communication is vital to uncover the causes of an unwanted behaviour. Advice sheet 1, in the Appendix, gives advice on structuring a behaviour consultation, and Advice sheet 2 provides a list of questions. Once the owner has answered these, you should have a better idea of what else you might want to know to identify the problem and come to a solution.

When taking your history, you will ask a number of questions of the owner. Internally, you should be trying to find answers to the following major points.

- *Do I think this problem is medical or behavioural in origin?*
- *How good is the husbandry of the rabbit and where is there potential for improvement?*
- *If there is more than one rabbit in the home, what is the relationship between the two rabbits?*

- *How much does the owner know about rabbit husbandry, and to what extent are they able to read their rabbit's behaviour?*
- *What does the owner see as the major problem here?*
- *Why do I think the rabbit is performing this behaviour?*
- *What type of solutions do I think will be adopted by the owner and improve the behaviour of the rabbit?*

Asking strategic questions will not only help you to get to a diagnosis more quickly, but also help you to develop an empathetic rapport with the owner. Identifying the owner's motivations and concerns will enable you to construct behaviour modification plans that can be actually achieved by the owner, and will help you jointly to agree on when to review. With any behavioural modification plan, there are many uncertainties, so being able to help set realistic expectations will increase the likelihood that the owner is satisfied with the outcome.

Asking about early history of the rabbit

When giving a history of the rabbit, many owners will place undue weight on the early life history of the rabbit ('We got her from a rescue centre, she had been neglected,' 'We bought her from a pet shop but I don't think they were looking after her very well,' or, 'We adopted him from a pet shop because he had grown too old to be sold'). While this information is interesting, if the rabbit has lived in a particular environment for several months, it will have adapted its behaviour to that environment. Therefore, a rabbit that shows unwanted behaviours more than a couple of months into a new home probably has deficiencies in its current health or welfare.

Nevertheless, socialisation is one example of an early factor that may affect behaviour later in life. In dogs, many behavioural problems are caused by poor socialisation to humans, dogs, and environmental stressors during a critical developmental period. Owners have lower expectations of rabbits as pets (they aren't expected to leave the home environment and meet unfamiliar rabbits and humans), so the effect of poor socialisation is not as marked.

Some information on how the rabbit was socialised as it was raised may help you to set a realistic prognosis. Different studies have found positive effects of handling or exposure to human scent in juvenile rabbits of different ages (from neonates to four-month-old animals [summarised in Bradbury and Dickens, 2016]). As rabbits should not be sold until they are ten weeks old, this means that juvenile rabbits should be handled or socialised by the breeder before they are old enough to be sold. It is not known what proportion of young rabbits receive this best-practice exposure. Proper socialisation of young rabbits may increase their tolerance of humans, and willingness to interact with us. However, poorly socialised rabbits will still make good pets providing their husbandry and owner interactions are appropriate.

It is rarely helpful or useful to place the blame for a rabbit's behaviour problem on its early life experiences, as it distracts the owner from the current situation. It may also reduce the owner's belief that they can succeed in managing the problem behaviour, and hence reduce their motivation to take ownership and put in the required amount of effort.

Asking about how the rabbit spends its time

This gives you an idea of the rabbit's day-to-day life, and how the owner views its normal behaviour.

Rabbits have a 'time budget' and an 'energy budget'. This is because they have a finite amount of time available to them, and a finite amount of energy to spend. Abnormal behaviours that take up a lot of time (such as over-grooming) will reduce the amount of time that the rabbit has to spend on normal behaviours. Abnormal behaviours that take a lot of energy or effort (such as fighting or aggression) will reduce the amount of energy that the rabbit has to spend on normal behaviours.

Different husbandry practices also affect how the rabbits spend their time. Rabbits fed on highly calorific concentrate foods spend less time eating, so may be more likely to be bored.

Time and energy budgets can be useful both to assess the rabbit's current situation, and to consider how the owner can alter these budgets to improve the rabbit's welfare.

Asking about signs of pain

Detecting pain is very important, both from an ethical viewpoint, and also to maximise the chances of successful resolution of an unwanted behaviour. When a rabbit is putting effort into coping with pain, then it has less mental reserve to learn new behaviours or adapt to new environments. It is important to diagnose and treat any pain before implementing a behaviour modification plan to minimise the rabbit's suffering and maximise the chance of successful modification.

Humans find it hard to instinctively understand rabbit behaviour, and this extends to understanding behaviour changes that indicate pain or disease. For this reason, any behavioural consultation should involve a full clinical examination of the rabbit – this will be covered later. In addition, any examiner (such as a vet or behaviourist) is unlikely to see the rabbit in its normal environment and, since rabbits are a prey species, stressful situations (such as a veterinary examination) make their display of pain-related behaviours less likely. For this reason, when taking your history, you should ask about signs of pain (explicitly or not) as well as signs of disease.

Your detailed understanding of normal and abnormal rabbit behaviour may allow you to pick up on signs of pain that the owner has missed, and this should be an important part of your assessment. However, with most animals, people who know the individual pet are much more likely to notice signs of pain. As well as the behavioural suppression in times of stress mentioned above, familiarity with the animal's baseline makes it easier to spot changes. Therefore, you need to ask suitable questions and be alert to descriptions of behaviours that may signify pain.

Behaviours, either seen or in the history, which indicate pain are listed below (Paul-Murphy, 2006). Bear in mind that the signs of pain are different depending on whether the pain is acute or chronic.

If a rabbit is in acute pain, there are certain behaviours that it may show, depending on the site of the pain:

- *Limping*
- *Rubbing or scratching at an area of the body*
- *Adopting an unusual posture (hunched or stretched out when not relaxed, Figure 2.1)*
- *Grinding teeth loudly (quieter tooth grinding is associated with contentment)*
- *Sitting with eyes partly closed*
- *Breathing faster*
- *Sitting very still and being reluctant to move*

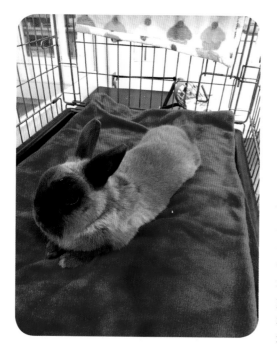

Figure 2.1: A stretched-out rabbit is not necessarily relaxed. This rabbit is in a kennel at a veterinary surgery.

Signs of acute pain are listed in the box. However, if you are assessing a rabbit with a behavioural problem, the owner is more likely to describe signs of chronic pain (unless the problem is linked to a sudden-onset medical problem).

Chronic pain in a rabbit is even harder to detect than acute pain. Chronic pain is associated with many common diseases: dental disease, chronic gastrointestinal disease, bladder stones, and arthritis, among others. Looking for changes in normal behaviour is useful. However, if the pain has been present for months, or even as long as the owner has had the rabbit, then they may not have a baseline benchmark with which to compare.

Case study 1: Pepper

Owners may not recognise signs of ill health in their rabbit. Pepper was a two-year-old male Mini Lop rabbit who was brought to the vet for vaccination. When asked why the rabbit was called Pepper, the owners said that he had sneezed frequently ever since they had acquired him at eight weeks of age. The owners had never thought that this might be a sign of ill health.

When an animal is in chronic pain, the motivation to avoid worsening the pain can be stronger than most other motivations. They may reduce their normal behaviours (like exploring, moving around, or eating). A rabbit that is in chronic pain is less likely to perform normal behaviours than is a 'normal rabbit'. A normal rabbit is bright, alert, active, inquisitive, with a good appetite, and it tries to get away from danger. Painful rabbits may behave differently.

These are examples of behavioural warning signs that you may hear when taking a history. A rabbit that is chronically painful may show any of these signs.

- *The rabbit has a reduced response to food. ('She used to snatch spinach out of my hands, but now I put it in her hutch and she just eats it during the day')*
- *The rabbit is reluctant to move. ('He's really lazy')*
- *The rabbit shows an increase in defensive behaviours. ('She had always been fine with me picking her up, but a few months ago, she started attacking me for no reason')*
- *The rabbit sits in postures that are unusual for a given situation. ('I don't think he's stressed, he spends*

The signs described in the box are not all specific for pain – they can also indicate fear or learned helplessness. Therefore you should consider these signs in the context of other information, especially any history of disease or abnormalities that you observe on the clinical examination.

Keeping a diary of the behaviours

Alongside a verbal history from the owner during the consultation, you can review a behaviour diary. In this, the owner records instances of the problem behaviour, together with information on the situation, triggers, and consequence. This often gives more reliable information than relying on an owner's recall, and can show trends or patterns to the behaviour. Advice sheet 3 gives an example behaviour diary. This can be a useful way of identifying the antecedent to the behaviour, the type of behaviour and the consequence to the animal and the owner.

Ask the owner to bring the diary to the consultation, and use this to inform the specific questions that you ask. This helps the owner to recognise patterns for themselves, so they may already be more amenable to the sort of changes that you're likely to suggest. It also

helps to set the owner's expectations of ongoing work – that he or she will need to work on the problem behaviour at home and then report back to you. The more involved that owners can feel in the process of diagnosis and treatment, the more likely they are to comply with the recommendations that you make.

Visiting the home

Home visits will show you the rabbit's normal environment, which helps you to diagnose the problem and make a treatment plan. If you have the opportunity, home visits are extremely useful. When they are reporting it to you verbally, owners may, consciously or unconsciously, give an overly positive description of their husbandry and interactions with their rabbits. However, if you can see the environment first-hand, you can calibrate what the owners say with what you observe. It enables you to see how the owner and rabbit interact. It also enables you to give very practical recommendations based on small details of the environment or interactions, and show them which specific behaviours of their rabbit indicate positive or negative emotional states (Figure 2.2).

However, home visits are by no means essential. Pragmatically, home visits are usually more expensive, and cost is usually a limiting factor in any behavioural work-up. Many rabbit behavioural problems are only mentioned to vets at vaccination or nail clipping consultations. Suggesting an expensive treatment course may prevent the owners from engaging with the treatment plan, and we must not discourage owners from seeking help.

While home visits are desirable in any behavioural work-up, you may still not see the true picture. Rabbits are very sensitive to changes in their environment, and your presence as an observer may affect the rabbit's behaviour.

Figure 2.2: A home visit allows you to see the rabbit when it is relaxed in its natural environment, and allows you to tailor your advice appropriately.

Owners may also change their behaviour when they know they are being watched. Nevertheless, if possible, try to see the owner and rabbit in the home environment.

Photographs and videos

On many occasions, home visits are not possible, and, even if you are able to visit the home, you may not see the unwanted behaviour. Asking the owner to take photos of the rabbits' home environment can be very useful. Additionally, this helps the owner to feel more involved in the treatment process.

Still photographs are indispensable for assessing the environment in which the owner keeps the rabbit. One photo should cover the usual living area (if possible), including sleeping area, exercise area and environmental enrichment. Others should show any other spaces to which the rabbits have access, how the owner gives food to the rabbits (food bowl, water bowl or bottle, hay rack) and where the rabbits urinate and defecate. Additionally, it is useful to see photos of where the rabbit chooses to spend time in the environment. Where does the rabbit like to sit? Where does it sleep? How does it sleep (Figure 2.3)?

Interactions and behaviours are best shown using video. If there are two rabbits, ask the owner to take a video of the rabbits interacting with each other. If there are any problems with the human–rabbit bond, then a video of the rabbit's response to a proffered hand or to being stroked will also provide useful information – both on how the owner attempts to initiate contact, and how the rabbit responds. Encourage the owner to film the unwanted behaviour – this is obviously easier when the owner has some understanding of what triggers the behaviour. Finally, if possible, encourage

Figure 2.3: Photographs may also show unusual postures of the rabbit, which may indicate pain or discomfort.

the owner to video a time where they think the rabbit is happy. This will give useful information on how well the owner understands the rabbit's behaviour.

Owners should be reminded to film the rabbits at floor level and not lift them on to a surface.

Clinical examination

The next important component of the diagnostic process for an unwanted behaviour should be a thorough clinical examination performed by a vet. Many problems that appear to be behavioural have an underlying medical cause.

Clinical examinations are not pleasant for rabbits, and you (whether you are the person examining the rabbit or the person bringing the rabbit in to be examined) can make the process more humane. Where possible, rabbits should be brought into the surgery in boxes with a

removable upper side (i.e. a lid the same area as the floor), so the top can be opened and most of the clinical examination can be performed without needing to lift the rabbit out of the box. Dragging a rabbit out of the front of a cage is stressful to the rabbit and increases the risk that the handler will get bitten.

If the rabbit has a companion, both rabbits should be transported to the vet (Figure 2.4). There has been some debate over whether the stress to the other rabbit is justified by the stress reduction in the rabbit that requires veterinary attention. However, in addition to the stress of transport, there is a stress arising from the separation of the rabbits, which will affect both animals. There is also the risk that the relationship will be disrupted when the first rabbit is brought back. On balance, it is sensible to keep a bonded pair of rabbits together as much as possible.

By the time the rabbit has arrived at the veterinary surgery, it will have become somewhat

Figure 2.4: If the rabbit is bonded to a companion, both rabbits should be transported in the same cage.

familiar with the box. Certainly the box will be perceived as being safer than the consultation room (both due to familiarity and the restricted space, which is preferred in times of danger). Leaving the rabbit in the box as long as possible will allow it to feel more secure. Additionally, gradually increasing the intensity of the fearful stimuli, rather than 'flooding' the rabbit with many fearful stimuli, will allow it to cope better with each new stimulus.

In practice, this means that the box should be placed on the floor of the consultation room while the history is taken (allowing the rabbit to acclimatise to the novel scents). When the clinical examination is required, the rabbit should be left in the box as much as possible, and the initial examination should be visual. Then the rabbit can be palpated and auscultated while in the box. Carrying out the entire clinical examination on the floor reduces handling of the rabbit, and prevents the rabbit slipping or jumping off the table and injuring itself.

The rabbit can then be lifted out of the box to be weighed, to be turned upside-down for assessment of the anogenital area and paws, and then can be restrained for claw clipping (if necessary). We must be aware that being handled by an unfamiliar person, being inverted and having the paws touched are all very distressing experiences for these animals.

Finally, if the rabbit is unusually confident, it can be placed on to the floor for assessment of locomotion and gait. However, in an unfamiliar environment, most rabbits will freeze. In most cases, if a problem of locomotion or gait is suspected, the owner should be asked to email a short video of the rabbit moving around. As soon as the clinical examination is complete, the rabbit should be returned to its box on the floor.

Veterinary surgeons can perform clinical examinations in a way that is less unpleasant for the rabbit but still complete.

The veterinary surgeon can assess several variables while the rabbit is in the box:

- *Body condition score: both visually and by palpation*
- *Muscle mass over lumbar spine: provides information on how active is the rabbit*
- *Eye, nose and ear health*
- *Coat condition*
- *Palpation for facial asymmetry, abscesses or tooth root protrusion. Movement of jaw to assess frontal and lateral deviation of mandible*
- *Auscultation of the heart, lungs and gastrointestinal system*
- *Palpation of the abdomen*

The veterinary surgeon can then lift the rabbit out of the box:

- *Weigh rabbit*
- *Check condition of abdominal/ano-genital fur*
- *Check hind paws for severe pododermatitis lesions (ulcers on the soles of the paws)*
- *Check claws*
- *Assess bone and joint health (if any abnormalities suspected from history or from clinical examination)*
- *Oral examination (if poor dental health suspected from history or from clinical examination)*

Section on Health at p. 52 contains information on the common medical problems that should be considered when a rabbit presents with a behavioural problem.

Reviewing changes and assessing success

You should be enthusiastic but realistic, when an owner or client approaches you for advice on a problem behaviour. Be enthusiastic about the prospect of trying to improve the situation, but give a guarded prognosis for full resolution of the behaviour. You need to set appropriate expectations – the process of modifying behaviour is not quick and in many cases, you may never get full resolution of a behaviour.

A major cause of this, as hinted in the previous section, is that owners can struggle to make the appropriate changes to address a problem behaviour in their rabbits. The owner may find it very difficult to change the aspects of husbandry or interaction that have led to the problem. It is often more acceptable to an owner to resolve a medical problem by administering a medication than it is for them to resolve a behavioural problem as the latter may require significant changes to his or her way of life.

Getting an owner 'on side' with you is therefore extremely important. Try to involve them, as much as possible, in the process of identifying the cause of the unwanted behaviour and working out how to resolve the behaviour. Owners who feel that they are an important part of the process are more likely to understand the rationale behind the recommendations that you make. They are more likely to feel sympathy for, rather than frustration with, the rabbit. This will increase their motivation to continue working until they see a change and, therefore, the likelihood of a successful resolution. Additionally, showing them videos of rabbits performing the desired behaviour will provide a vision of what they could achieve and prevent them from losing hope.

When giving advice on changing behaviour, avoid constructing a behaviour plan, and then turning over the entire responsibility to the owner. Progress may be slow and results may not become apparent for several weeks. This is long enough for owners to believe that the intervention has not worked and to give up. Therefore, there is a lot of value in planning reviews with the owners and describing what success would look like and how much progress they should expect at each review.

Ask the owner how they would assess success. They may be satisfied with a rabbit that simply doesn't perform the problem behaviour. Or they might be hoping for an outcome that is more nebulous.

'What is your goal from this behaviour consultation? What would be a good outcome for you?'

'I want my rabbit to like me.'
'I want my rabbit to sit on my lap.'
'I want my rabbit to stop biting me.'

You need to understand what the owner would like. In many cases, you will be able to help them to clarify exactly what they mean, which will help them to recognise and appreciate positive changes in their rabbit's behaviour.

'When you say that you want your rabbit to like you, how would you judge this? What behaviours would you like him to show?'

You may also be able to moderate unrealistic expectations, so you can prevent the owner from seeing a more moderate change as a failure.

'There are very few rabbits who like to spend a lot of time sitting on a human's lap, because they often feel quite constrained, and they haven't got a secure surface for their paws. Many owners say that they enjoy sitting on the floor and stroking the rabbit while they're watching television. That way the rabbit can feel secure and really enjoy spending time with the owner. It might be easier to aim for something like this in the first instance, as it is much more achievable.'

Once you are aligned on the end goal of the behavioural interventions, it's worth setting up when you'll next discuss how the rabbit is behaving. Make sure that you are clear that you're not checking up on how the owner is doing, but you're seeing how things are going and answering questions as they occur. This also helps you to ensure that the prognosis is realistic.

'With these sort of unwanted behaviours, it can take quite a few weeks before we start to see good progress. As every rabbit is different, I usually arrange a time to speak with the owners after about two weeks, so we can see what is working and what isn't working, and ensure that we adapt what we're doing so we get success. Would this be OK with you?'

Once the review meeting is arranged, be clear about what you want to assess at that point.

> *'When we have a chat in two weeks' time, I'll ask you how often the rabbit has thumped his back feet when you are close to his enclosure. I think you might start to see a few different behaviours when you approach the run, so if you notice them, make a note and we'll talk about them in two weeks' time.'*

These caveats and follow-up discussions maximise the likelihood both that the behaviour modification plan is successful, and that the owner is satisfied.

3

Why is the problem behaviour occurring?

This text has so far described how to take a detailed history and examine the rabbit for underlying medical problems. The next step is to collate the information gathered on how the rabbit is kept, compare it to what the rabbit needs, and then identify where there are discrepancies that may be affecting the rabbit's behaviour.

The state of an animal as it attempts to cope with its surroundings is called its 'welfare'. Animals can have good welfare or bad welfare. An animal with good welfare should experience comfort, pleasure, interest, confidence and a sense of control. To ensure good welfare, the animal's physical and emotional needs must be met.

Abnormal behaviours may be both a consequence and a cause of poor welfare. If a rabbit cannot express normal behaviours, or is suffering from poor husbandry, it may show abnormal behaviours. Conversely, if a rabbit expresses normal behaviours that are not desired by the owner, the human–animal bond may be disrupted, and this may negatively affect its welfare.

When looking at any unwanted behaviour, it is important to first assess the animal's welfare. What aspects of the welfare are good? What aspects can be improved? In what areas are the animal's needs not being met? If the owner's expectations are in harmony with the rabbit's requirements, and the husbandry is suitably designed, the risk of behavioural problems is very low.

We can think about a rabbit's welfare in various different ways. Many pet owners will be familiar with the 'Five Freedoms' model for assessing animal welfare, which forms the basis of the Animal Welfare Act (2006).

The Five Freedoms model was developed in 1965 as part of a government report on livestock welfare. It was rolled out in a 1979 press statement from the Farm Animal Welfare Council (FAWC) (2009). The report lays the Five Freedoms out as listed in Table 3.1.

> The FAWC released the following statement: 'We believe that an animal's welfare, whether on farm, in transit, at market or at a place of slaughter should be considered in terms of 'five freedoms'. These freedoms define ideal states rather than standards for acceptable welfare. They form a logical and comprehensive framework for analysis of welfare within any system together with the steps and compromises necessary to safeguard and improve welfare within the proper constraints of an effective livestock industry.'

While this model forms the basis of current animal welfare legislation in the UK, there have been some problems with its interpretation (Mellor 2016). The quotation above states that these freedoms define 'ideal states rather than

Table 3.1 The Five Freedoms as described by the Farm Animal Welfare Council.

Freedoms	Provisions
1. Freedom from thirst, hunger and malnutrition	By providing ready access to fresh water and a diet to maintain full health and vigour
2. Freedom from discomfort and exposure	By providing an appropriate environment including shelter and a comfortable resting area
3. Freedom from pain, injury and disease	By prevention or rapid diagnosis and treatment
4. Freedom from fear and distress	By ensuring conditions and treatment which avoid mental suffering
5. Freedom to express normal behaviour	By providing sufficient space, proper facilities and company of the animal's own kind

standards for acceptable welfare', but this point is often lost when the Five Freedoms are used. It is clear that, if all were held to be absolutes, then the animal would quickly run into problems: an animal that never experiences hunger is likely to be obese; an animal that never experiences pain would not know when it injured itself.

This model also implied that a caregiver would be under obligation to keep animals completely free of these negative experiences or states at all times. Much as behaviour can produce positive and negative 'effects', it can be produced by positive and negative 'affects'. Negative affects (unpleasant feelings or emotions), such as hunger and pain, are vital components of behavioural mechanisms that ensure the animal survives. For example, minor skin discomfort stimulates grooming. So we should not aim to completely eliminate them, but we should minimise the extremes of these negative affects. They motivate behaviours that keep the rabbits alive.

Additionally, the Five Freedoms don't really address positive experiences for the animal: providing opportunities for positive experiences is at least as important as minimising negative experiences. Providing animals with spacious, stimulating and safe environments allows them to perform rewarding behaviour such as exploration and foraging for food; providing animals with suitable companionship allows them to form good social relationships, which are rewarding.

Animal welfare scientists are starting to model animal welfare using a different framework, the 'Five Domains Framework', and we can apply to pet rabbits (Diagram 3.1). The 'Five Domains' model considers how the animal's physical state, environment and ability to express normal behaviours affect its 'affective states' (mental status, comprising subjective feelings or emotions). There are two types of sensory input that give rise to the 'affective state' of a rabbit. The first form of sensory input arises from the imbalance or disruption in the physical or functional state of the rabbit: hunger and thirst, weakness, pain and discomfort. The second form of sensory input contributes to the rabbit's cognitive assessment of its situation: anxiety, fear, frustration, helplessness, loneliness, boredom and depression. These two forms of sensory input influence the 'affective state' of the rabbit, and that relates to its welfare.

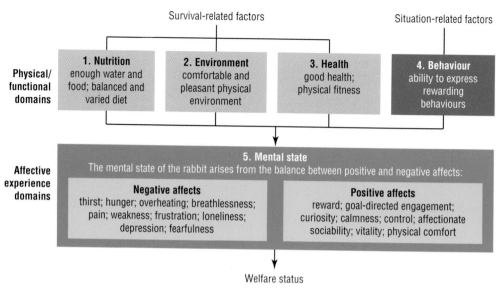

Diagram 3.1: The Five Domains framework for assessing rabbit welfare.

There are two ways to improve welfare. We must reduce negative affects to low levels, and we must provide environments that give animals opportunities to experience positive affects.

Physical/functional domains: Survival-related factors

The 'Five Domains' include four physical or functional domains, which relate to the physical state of the rabbit. Three of these are so-called 'survival-related factors': failure to meet the rabbit's requirements in these domains may cause death. These domains are nutrition, environment and health. As explained, these alter the 'affect' of the rabbit by altering the sensory input from its internal physical or functional state. When we assess a rabbit, we may see negative aspects (for example, restricted food and water, inappropriate diet) or positive aspects (enough food and water, balanced diet) within each domain.

In this section, we'll describe the requirements of the rabbit in each domain, and we'll explore how different husbandry practices may result in negative or positive affects.

Nutrition

We'll start with rabbit nutrition – a subject that has been the focus of a lot of excellent research recently. When thinking about rabbit nutrition, the fundamental principle is that rabbits have evolved to eat grass.

Wild rabbits spend between 11 and 13 hours above ground each day, and spend between 30% and 70% of this time grazing.

The optimal way to feed rabbits would be to allow them to graze on a large area of grassland throughout the year. During the spring and summer, the higher carbohydrate levels in the grass and the increased grass growth will provide more calories. This creates a seasonal variation in the body condition score (this relates to level of body fat) of the rabbits.

Figure 3.1: In the summer, grass is lush and abundant, and in the winter, it is of much poorer quality. This seasonal variation is important for the metabolic health of many herbivorous species.

In many herbivorous animals, this rhythm of weight gain and weight loss is important to maintain insulin sensitivity and prevent metabolic syndrome: this has not been demonstrated in rabbits, but the natural diet of rabbits includes periods of higher- and lower-quality food (Figure 3.1).

The amount of land that will produce sufficient grass varies with climate, soil and amount of light, but as an example, a 10 m × 10 m lawn in partial shade in southern England produces sufficient grass for two medium-sized rabbits. The grass will overgrow during the summer and have sufficient plant mass to sustain the rabbits over winter. Practically, most rabbit owners do not have access to a large enough area of grassland to allow the rabbits to graze all year round. In addition, concerns about risk of predation (rightly or wrongly), and expectations of interactions with pet rabbits will make this method of feeding very uncommon. If rabbits cannot graze all year round, then substitute foods must be provided.

Types of rabbit feedstuffs

The common types of rabbit food are as follows:

Forage feedstuffs

'Forage' is the term given to plant-based feedstuffs that are high in cellulose (i.e. leaves and stems, as compared to sugary fruit or starchy vegetables). For rabbits, this is typically grass or hay.

Grass

We use the word 'grass' to describe the leaves and stalks of grass species (monocotyledon plants, where the seed yields just one embryonic leaf). Grasses typically contain high levels of silica, to deter herbivores from eating them higher silica levels reduce grazing by some herbivores, and also reduce the ability of the herbivores to absorb nitrogen. Additionally, there is dynamic feedback: grasses that are grazed increase the silica level in their leaves. To deal with this, rabbits have evolved teeth that grow continuously. If they did not, the teeth would quickly wear down and the rabbit would be unable to chew food.

Grass has a high water content, low energy density and is very palatable to rabbits. Grazing is a rewarding behaviour. Other food sources can provide some of the benefits of grass, and they all differ in which requirements they can meet and which they cannot.

Hay

The 2017 PAW report found that 33% of rabbits were not being fed enough hay. Significantly more female rabbit owners are feeding the right amount of hay (53%) than male owners (38%). Rabbit owners who are familiar with the Animal Welfare Act are more likely to feed the right amount of hay (60%) than owners who are unfamiliar with it (39%).

Figure 3.2: Good quality hay has long pieces of different grasses, and smells sweet.

Hay is dried grass. It is typically a mixture of fescue, ryegrass, timothy, brome and cocksfoot grasses. Good quality hay has a mixture of grass types, the pieces of hay are longer than 10 cm (i.e. it is not finely chopped) and it smells sweet and pleasant, not musty or mouldy. Good hay is, at least partly, green. (Figure 3.2).

Hay can be bought from pet shops or horse feed stores. Much of the hay produced for small pets is finely chopped so it is easier to pack into small bags. However, this makes it far less appealing for rabbits. Rabbits interact with fresh hay in a variety of ways: they nip off stalks that are sticking out from the clump (as they do when grazing, which stimulates more grass leaf growth and delays the plant from going to seed), they often use their paws or nose to nudge the hay into a different shape and they select choice pieces of hay to eat. If rabbits have the opportunity to express these different behaviours, they are more likely to interact with the hay, and therefore more likely to eat it (Figure 3.3). This is important when encouraging rabbits to eat a more hay-based diet. Owners are more likely to increase the hay consumption of their rabbits if they provide long-chop rather than short-chop hay. Additionally, hay intended for horses usually has a higher turnover rate (i.e. less time on the shelf or in the store) than hay for small animals, so it is likely to be fresher. Owners who buy a whole bale of hay are more likely to feed large quantities to their rabbits daily, because it is much cheaper, and there is much more available. This is better for the rabbits.

If owners cannot store a hay bale so it stays dry and fresh (the author uses a wheelie bin), then they will have to use a product for small animals. If this is the only option, you should encourage them to pick plenty of long grass, to ensure that their rabbits consume sufficient long fibre to optimise their health.

Figure 3.3: Long-chop hay is more likely to stimulate rabbits to interact with it, which increases the likelihood that they will eat it.

Straw

Straw is the dried stems and leaves of commercial grain crops, such as wheat or barley (i.e. a by-product of grain production). It may be yellow or brown. The stalks in straw are much tougher and thicker, and the leaf-to-stalk ratio is lower (the leaves of wheat and barley are largest in spring, and shrink as the grains reach maturity, which is when the straw is harvested). When a rabbit grazes, it selects pieces of different grass or specific plants, but straw comes from a crop, and so is all one species, which makes this behaviour harder to perform. Straw is less palatable than hay. Therefore, plenty of hay (not straw) should be provided to ensure that the rabbits eat enough. Owners are often confused about the difference between hay and straw, so you should ask them to describe the product if they are not sure which it is.

Concentrate feedstuffs (muesli mixes or monocomponent feeds)

As described in the previous paragraph, rabbits select choice pieces of food to eat. This is not a problem if the selection of food available comprises different species of grass. However, if the food contains some elements that are very palatable, but cause disease, then selective feeding will damage the rabbit's health. This problem is made worse when rabbits are fed ad libitum (as much as they want) – the rabbits will only eat the palatable, inappropriate pieces.

Pelleted rabbit feed is monocomponent: every piece looks the same. Fifty years ago, these pellets were sold as a cheap way to fatten rabbits for meat. Muesli mixes were then introduced to offer variety to the rabbits: these were perceived as a higher-end, more welfare-friendly product. Unfortunately, as further research was carried out into rabbit nutritional needs, it was found that muesli mixes encouraged selective feeding and caused a variety of diseases. There are now many options for concentrate foods for rabbits: these are all monocomponent, so do not permit selective feeding.

All concentrate foods have similar problems. They are designed to be palatable (so have higher levels of sugars and starches than does grass, which encourages overconsumption), they are designed to be stored for a long time (so have a very low water content, which does not support good urinary health) and they provide more calories for the same amount of chewing as hay (so increase the risk of dental disease). As hay is an inexpensive commodity (so doesn't make much profit), pet food companies are motivated to make a 'complete' diet for rabbits that they can sell for a higher price.

Concentrate foods should be fed as a treat. The majority of the diet should always be grass or hay.

Muesli mixes

Muesli mixes typically contain flaked peas and maize, coloured starch-based biscuits, whole or rolled grains (wheat and oats) and compressed pellets (Figure 3.4), which are designed to 'balance' the rest of the mixture (i.e. fats, proteins, vitamins and minerals). Some also contain chopped straw or hay.

> The 2017 PAW report found that 25% of rabbits were still being fed on muesli mixes. In 2016, 85% of vets believed that rabbit muesli food should be completely removed from sale.

Unfortunately, the components of the muesli mixes are very unlike the grass that rabbits have evolved to eat. They contain high levels of starch and sugar, which favour growth of disease-causing

Figure 3.4: Rabbit muesli mixes are more appealing to the owner and palatable for the rabbit, but cause significant health and welfare problems.

microbial species in the gut. Additionally, the different components of the muesli mix differ in palatability: the most palatable parts are eaten preferentially (the starchy flaked peas and corn) and the less palatable pellets are left. This means that the mix consumed by the rabbit is not 'balanced' as the manufacturer intended.

When rabbits eat muesli mix for long periods, various health and welfare problems become apparent. Muesli-fed rabbits are much more prone to dental disease: grains do not contain enough silica to wear down the teeth, and they do not require much chewing before they are swallowed. Muesli-fed rabbits are much more likely to be obese: the foods are much more calorific and are easier to eat. They are more likely to suffer gastrointestinal problems: high-sugar foods promote the growth of unhealthy bacteria, which change the acidity of the gut by producing lactic acid rather than the volatile fatty acids that are important for gut health – this also reduces the rabbit's motivation to eat the caecotrophs it produces. Muesli-fed rabbits are more likely to show behavioural problems: both from increased time budget (as they are eating less forage) and perhaps from the intestinal discomfort caused by the inappropriate diet. There has been some excellent work done on this at the Royal (Dick) School of Veterinary Studies in Edinburgh (Prebble et al., 2015). Studies into the effects of muesli mixes on rabbits have found that, even if rabbits eat all of the muesli mix, they still can't get all of their nutritional requirements from this feed. More forage feed, like hay and grass, is necessary for good health. However, even when muesli mix is fed with ad libitum hay, the rabbit still suffers adverse health and welfare problems: obesity, dental disease, behavioural problems, gut problems and urinary problems. While muesli food is very palatable, the adverse effect on health and behaviour means that it should never be fed to rabbits.

I will repeat that: muesli food should never be fed to rabbits.

Monocomponent feedstuffs

Monocomponent concentrate feedstuffs contain the same ingredients in every piece (Figure 3.5). Additionally, rabbits in the wild will eat many different types of plants, so limiting them to just one food will lead to boredom. These ingredients range from ground-up muesli, which contains high levels of starches and sugars, but at least prevents selective feeding, to grass meal, which contains high levels of cellulose but does not require the same amount of chewing that grass or hay does, so may lead to dental disease. Therefore, these foods should never be fed without forage, despite manufacturers' claims that the food is 'complete'.

While monocomponent foods prevent selective feeding, the ingredients can differ substantially. Cheap feeds contain the same ingredients as muesli (grains, legumes, starches, sugars, and oils), but these are ground and compressed. More expensive feeds often have higher levels of ground cellulose.

There are two processing methods used to produce rabbit feed – pelleting (where the food is ground and compressed, and the resultant pieces look shiny) and extrusion (where the food is heated and extruded from a die – the resultant pieces look like kibble). Early rabbit foods were pellets, and these were perceived as a cheap way to fatten rabbits for meat. Therefore, they have retained their low cost because they have a bad reputation with rabbit owners. Extruded feed then became fashionable. However, it is easier to incorporate higher levels of fibre into pellets than it is to put this into extruded nuggets, and the pelleting process can be done at a smaller scale than the extrusion process, so pellet preparations may become more premium again.

Figure 3.5: Monocomponent extruded foods are palatable and prevent selective feeding, but should never be fed ad libitum.

In dogs, Sechi et al. (2017) reported that commercial food preparations, which comprise carbohydrate, fat and protein sources that are not those commonly encountered by feral dogs and wolves, lead to adverse changes in neurotransmitter levels and increase the risk of behavioural problems. When managing behavioural problems in rabbits, it makes sense to transition them on to a diet that is as close to the wild diet as possible (much easier for rabbits than dogs!): such a diet does not include concentrate food.

COMPRESSED 'PELLETS' Manufacturing pet foods by pelleting is cheaper than extruding the feed. This means that the manufacturing method is more frequently used for lines with cheaper ingredients: pellet foods are usually of lower quality than nuggets.

EXTRUDED 'NUGGETS' Nuggets tend to be a more premium product, usually containing grass or grass meal as the major ingredient. Nuggets are a useful treat food for training, and good as a supplement to a grass- or hay-based diet to give small quantities of minerals.

CUBES Rabbit food cubes are sold as a complete food, and consist of timothy and alfalfa hay bound with cereal meal and soya oil. Some rabbits like them, but the lack of protruding strands reduces the motivation to engage with the product. The cereal meal and oils increase the calories of each mouthful, reducing the time that the rabbits spend chewing as compared to hay. Food cubes are sold as a convenient (though expensive) way for owners to provide plenty of forage, but are unlikely to provide as much behavioural stimulation for rabbits as would long-chop grass or hay, and so they should always be fed alongside ad libitum hay for good behavioural health (despite the optimistic claims of the manufacturer!).

STICKS 'Stick' food for rabbits describes chopped hay bound with cereal meals and soya oil, which is then extruded into thick 'sticks' of around 10 cm in length. As with the cubes, the processing of this product removes protruding strands, and the inclusion of cereal meals and oils increases the calorific value. In order to extrude these sticks, the length of the hay within the sticks is short (a few centimetres), which reduces chewing time. As with cubes, rabbit 'sticks' should always be fed alongside ad libitum hay – it is crucial that concentrate foods are not fed as the major portion of the diet.

'TREATS' There are a wide variety of commercially available rabbit treats. These are typically high in starchy carbohydrates and oils and low in fibre. Many are described as 'natural' or 'healthy', and include small quantities of ingredients that are found in the rabbit's natural diet (Figure 3.6). They are labelled as 'treats' or 'complementary food', so they have no requirement to meet the rabbit's dietary needs. Owners should be advised not to feed such products as, although palatable, they may lead to gastrointestinal problems or dental disease and are not suitable food for rabbits.

One suggestion of a good diet for rabbits is shown in Diagram 3.2. Green leafy plants provide more variety. Concentrate foods are optional, so many owners choose to feed 15 per cent green leafy plants and rarely give concentrate food at all.

Owner misconceptions around rabbit nutrition

There are several reasons why it may be difficult to persuade owners to feed rabbits in a way that

Figure 3.6: Commercially available rabbit 'treats' are not suitable foodstuffs for rabbits. Owners should be advised to feed 'treats' of leafy greens or small quantities of monocomponent food.

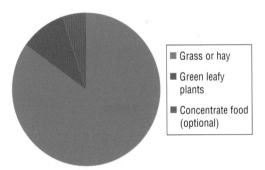

- Grass or hay
- Green leafy plants
- Concentrate food (optional)

Diagram 3.2: A good diet for rabbits has these proportions of different foodstuffs.

is best for the health and welfare of the rabbit. As humans, we often resort to anthropomorphism when we don't know about an animal's needs – projecting human wants and needs on to the animal. This tendency to anthropomorphise results in several problems around rabbit diets. These are listed below, and then each point is explored in detail.

- Humans eat meals, so think that rabbits should eat meals.
- Humans like to eat a variety of food, so think that rabbits would want to eat a variety of food.
- Humans can't eat grass, so think that grass cannot form the major part of the rabbit's diet.
- Humans often don't like to eat green vegetables, so think that 'greens' are not a treat.
- Humans don't eat their bedding material, so don't think rabbits would want to eat their bedding material.
- Humans eat food from bowls, so think that rabbits should eat food from bowls.
- Humans don't have continuously growing teeth, so don't understand how they work.
- Many humans equate spending money with providing care, so if food is 'free', think that they are not providing the best care for the rabbits.
- Humans think that bottles of water are cleaner than bowls, but don't realise that they are harder to drink from.

Humans eat meals, so think that rabbits should eat meals.

Humans define a meal as a portion of food that is provided intermittently and consumed when it is provided. Rabbits are evolved to eat almost continuously, pausing only for rest or social interactions – this eating pattern differs from both humans and the other commonly kept pet species. Dogs tend to eat intermittent large meals, as befits an animal that hunts or scavenges. Cats tend to eat many small meals during the day, as though catching small rodents. Humans usually eat a few meals per day. Therefore, many owners do not realise

that the rabbit's eating pattern is different to their own.

When given access to large areas of grass, rabbits will eat leisurely, spending the majority of their time grazing. However, if a rabbit is given a very palatable food (i.e. one for which it is not evolved), it will consume it very quickly. When pet owners provide a novel food to the animal, they judge the food on how quickly the animal approaches the food and how quickly it consumes the food, rather than how closely the food mimics the animal's natural diet. Concentrate foods (both monocomponent foods and muesli mixes) are designed to be extremely palatable, and consequently are often eaten quickly by the rabbit, fulfilling the owner's desire to provide a 'mealtime' experience for their animal. This may become one of the major ways that the owner interacts with the rabbit, so even if it is damaging to the rabbit, it can be hard for the owner to change this behaviour without feeling that they are damaging their relationship with their pet.

Humans like to eat a variety of food, so think that rabbits would want to eat a variety of food.

Rabbits, like humans, do like to eat a variety of foods. They have a sweet tooth, but high levels of sugars and starches cause disease. The variety should come from different grass and leafy plant species.

As omnivores with a poorly developed sense of smell, humans do not recognise that there is substantial variation in odour, texture and taste of different grass and leafy plant species. Rabbits are selective feeders, and they will make choices during grazing. In a large area, rabbits may selectively eat different leafy plant species and a range of grass species. In order for rabbits to differentiate between poisonous and edible plants, they have developed a sophisticated sense of taste. Humans have about 10,000 taste buds on the tongue, but rabbits have around 17,000. Rabbits are able to discriminate between many more flavours, and can detect those flavours at lower concentrations.

Like humans, rabbits do have a sweet tooth. Rabbits will opportunistically eat fruit if it falls from the trees. However, since they cannot climb up to get it, most fruit will be taken by other animals that can. This means rabbits have not evolved to cope with high levels of sugar in their diet. Plants that survive being grazed by rabbits (i.e. grasses) spread by runners as well as by seeding – plants that are heavily grazed do not produce much grain. Rabbits have also not evolved, therefore, to eat much food that, like grain, is high in oils or starch. Evolution or not, rabbits still find both high-sugar and high-fat/starch foods palatable, although they are not good for the animal's health.

This tendency to selectively feed on high-sugar and high-starch foods means that rabbits rarely consume all of a muesli mix. There has been a concerted education campaign on the problems with muesli feeding from various rabbit welfare charities, reinforced by refusal of a major pet shop chain to sell muesli food, and this has caused a significant reduction in the rates of owners feeding this food. It has been a difficult habit to change: the visual appeal of muesli (consisting of grains, flaked peas and corn, dried fruit, cereal biscuits and pellets) appears much more appetising to an owner than does a brown homogeneous monocomponent food (the appearance of food does not matter to the rabbit, as the position of their eyes means that they cannot see under their nose).

Owners can still give a variety of foods and 'treats'. However, they need to provide the sort of variety and types of foods that the rabbits would encounter in the wild.

Humans can't eat grass, so think that grass cannot form the major part of the rabbit's diet.

Rabbits, unlike humans, can extract nutrition from cellulose, the major component of grass. Humans can't digest grass, so we do not intuitively understand it is food. For this reason, it can be hard to persuade owners to feed sufficient forage and reduce or even eliminate concentrate foods: it makes them feel that they are in some way neglecting their pet. Humans do not have a frame of reference for the taste of different grass species, so don't necessarily believe that grass can taste good.

Concentrate foods (any commercial 'rabbit foods') are designed to be very palatable so the owners continue to buy the food for the rabbits. In addition, it is in the interests of the pet food companies to encourage owners to feed more food rather than less. A study in Edinburgh found that feeding only hay to rabbits did not cause ill health (Prebble and Meredith, 2014), and this diet allowed them to express normal behaviour. For adult rabbits, owners should consider concentrate foods as an optional extra: useful for trick training and encouraging wanted behaviours, but not something to feed every day, and certainly not something to leave available all the time.

Humans often don't like to eat green vegetables, so think that 'greens' are not a treat.

Rabbits consume both grass and leafy green plants in the wild. However, the composition of wild plants differs from that of farmed green vegetables. Vegetables domesticated for human consumption have been selected for lower levels of bitter compounds (these compounds confer health benefits, but make the vegetable less palatable), lower levels of lignin (so the vegetables are softer) and higher levels of sugars (so the vegetables are more palatable).

This is the reason that rabbit dietary guidelines recommend restriction of vegetables – i.e. that they are fed as a 'treat' rather than an essential component of the rabbit's diet.

Owners should provide small quantities of fresh green leaves or vegetables as part of the normal diet. Leafy wild plants (dandelion, rose leaves, leaves on branches from fruit trees) are preferable to vegetables grown for human consumption. If owners wish, they can provide small pieces of vegetable or fruit as training rewards: pieces of carrot peel or apple skin about half the size of the fingernail on the little finger.

Humans don't eat their bedding material, so don't think rabbits would want to eat their bedding material.

Rabbits in the wild like to eat while defecating, and then to move on to a new grazing area when the grass is soiled (when the grass is refreshed by rain or new growth, the rabbits will eat it again).

Pet rabbits should be given every chance to express normal feeding behaviours, as this will increase their consumption of hay and grass, which is important for their health. As they have less space to move away from soiled hay, the owners need to provide fresh hay every day. When owners are paying for hay, their motivation is to ensure that all the hay is consumed or used so they don't waste money. This poses a challenge when encouraging owners to provide more hay than the rabbits can possibly eat: if the owner sees that the hay is still present, they may perceive that the rabbits have enough food. However, the rabbits see that the hay is soiled and not fit for consumption (Figure 3.7).

Owners should be advised to provide hay daily on the floor of the living quarters of the rabbit, regardless of the hay that is left. Research has found that rabbits prefer to eat hay from the

Figure 3.7: Rabbits frequently eat, defecate and sleep on the hay provided, so the owners must provide fresh hay daily.

Figure 3.8: Hay racks are popular ways of providing hay for rabbits, but rabbits prefer to eat hay from the ground. Owners should provide hay in a number of different ways to encourage the rabbits to eat plenty of hay.

floor as though they were grazing, rather than pulling hay from a hay rack (Prebble et al., 2015; Figure 3.8).

Humans eat food from bowls, so think that rabbits should eat food from bowls.

In the wild, there is rarely competition for food resources: the food takes time to consume and grows across a large area. For a pair of rabbits, humans often feed high-value foods (such as concentrate food or vegetables) in a single bowl. This causes unnecessary conflict and can lead to fights between bonded pairs (Figure 3.9). Reassure the owner that it is perfectly reasonable to scatter the food into the bedding (see previous point) as the rabbits will find it all, it won't be soiled (unless the cage is very dirty: when it should be cleaned) and the rabbits will spend more time foraging (which alleviates boredom). Nearly all food-related aggression is caused by inappropriate methods of feeding.

Humans don't have continuously growing teeth, so don't understand how they work.

Rabbit teeth grow continuously throughout life – ensuring that the teeth don't wear away (due to high-silica grasses) and that the rabbit is always able to eat. This concept is rather alien to us, as our teeth stop growing in adolescence. It is important to educate owners that the teeth are evolved to wear down at the right speed if rabbits are fed on the foods that they are evolved to eat.

Some owners will provide various items for the rabbit's teeth: unpalatable dry sticks (which they rarely chew on), abrasive toys (which rabbits do not chew with their molars) and even foods that are perceived as 'hard', such as raw potato or dry bread. Rabbits are not motivated to wear down their teeth for the sake of their teeth, so these strategies are doomed to fail. Explaining that high-silica forage (i.e. grass or

Figure 3.9: Feeding from a bowl increases the rabbit's speed of eating and decreases expression of normal feeding behaviour. Encourage the owner to scatter food in the enclosure or in the hay.

hay) is the only effective way of achieving this normal, and critical, wear is very important.

Many people equate spending money with providing care, so if food is 'free', think that they are not providing the best care for their rabbits.

The common domestic pet species are rarely fed on homemade diets: their dietary requirements are very hard to meet. For most pets, owners have to pay for their animal's food. By contrast, many rabbit owners have grass in their gardens, or can pick it from areas very close to their home – a free source of excellent food. Many owners believe that this grass must be in some way inferior to foods that they have bought specifically for their pet.

The benefits of feeding grass to rabbits, as previously discussed, outweigh the small risks. Freshly picked grass has a higher water content than hay, promoting more urine production

and reducing the risk of bladder stones or sludge (see section on health). Hay is an excellent substitute when grass isn't available: either because the owner does not have access to grass, or during winter when grass growth is low. However, hay is less palatable than grass, so owners may find it harder to encourage rabbits to eat sufficient hay to meet their needs without restricting concentrate foods. Freshly picked grass is almost always better for rabbit health than purchased hay.

Owners usually voice three major concerns about feeding handpicked grass: the risk of disease transmission from wild rabbits, the risk of diarrhoea and the risk of disease from fermenting lawnmower clippings.

There is some small risk of disease transmission from wild rabbits: however, if the pet rabbits are vaccinated and the owners do not live in an area that has a large population of wild rabbits, then the risk is very low, and, in my opinion, outweighed by the health and behavioural benefits of feeding grass. Some owners feel that they need to wash all the grass that they provide to their rabbit, which is time-consuming, messy and impractical, so this can discourage them. This is not necessary if the area of grass is rarely or never grazed by wild rabbits. Some soil on the grass may well be beneficial: the silica in the soil provides extra abrasion for the teeth.

Any dietary change can cause diarrhoea or gastrointestinal upset in a rabbit, but owners can reduce the risk of this by introducing the new feedstuff gradually.

Lawnmower clippings can cause diarrhoea in rabbits – the bruising and heat damage to the grass causes rapid fermentation, which can occasionally create toxic by-products. Therefore, it is correct that owners should not use lawnmower clippings as rabbit food. However, this rationale for not feeding lawnmower clippings to rabbits is rarely understood, and many owners think that this is due to some problem with grass itself.

It's important to recognise these underlying beliefs or motivations when trying to change an owner's behaviour. Owners need to believe that, despite being free, a diet based on grass and hay is the most natural, healthy diet that rabbits can eat: feeding anything else raises the risk of health or behavioural problems.

Humans think that bottles of water are cleaner than bowls, but don't realise that they are harder to drink from.
Rabbits have a higher water requirement for their size than humans. Failure to drink enough water predisposes rabbits to urinary tract disease and bladder stones. Traditionally, rabbits were given water from a bottle with a dependent ball valve, which the rabbits moved with their tongue and lips to get water. These bottles are convenient for the owners: they are less likely to be soiled by urine or faeces, or filled with hay, and so they appear to be cleaner for longer. Unfortunately, this means that owners tend to provide clean water less frequently, and (normally invisible) bacterial overgrowth is common. Additionally, water bottles were designed for the rabbit or small pet kept in a very small space. If rabbits have large enclosures, there should be sufficient space to provide a wide, stable bowl rather than a bottle (Figure 3.10).

A recent study showed that rabbits drink less from bottles than bowls: presumably because the behaviour is not instinctive and there is a higher time and energy 'cost' for the rabbits to access the water (Tschudin et al., 2011). Owners should always provide water in a bowl, even if they choose to provide a bottle as well (Figure 3.11).

Advice sheet 4 in the Appendix helps to guide owners through the process of transitioning their rabbits on to a diet that will help both their behaviour and their welfare more generally.

Puzzle feeding
'Puzzle feeding' refers to any way of providing food to an animal in which the animal needs to do work to acquire its food. This feeding method has become increasingly popular with pet owners because it has a number of benefits: it reduces aggression around a food bowl, it provides environmental enrichment and it slows down consumption. A further benefit may be that it increases the owner's interest in their rabbit's behaviour.

For rabbits, puzzle feeding is usually used to deliver nuggets: these are of a consistent size and convenient to put in the feeder. Also, they are highly calorific and need to be restricted, making puzzle feeding ideal. A popular method of puzzle feeding uses a ball

Figure 3.10: Bowls should be wide, stable and sufficiently filled so the rabbits can drink comfortably.

Figure 3.11: Some rabbits have learned to drink out of bottles: owners should be advised to offer a bowl alongside the bottle. Rabbits drink less from bottles than from bowls, and there have been reports of rabbits becoming severely dehydrated when the bottle valve fails.

that infrequently dispenses nuggets when rolled around (Figure 3.12).

While this can be a way of enriching the rabbit's captive environment, wild rabbits do not frequently need to work for food. For this reason, puzzle feeding may have less value than it does for dogs and cats, which would naturally hunt for their food. Certainly, it cannot replace access to grass to allow normal feeding behaviour. Nevertheless, puzzle feeding can be stimulating and interesting, and provides a good way for owners to interact with their pets.

An extension of puzzle feeding is training. Rabbits can (and will) learn to perform a specific behaviour, on command, for a food reward. Rabbits can learn multiple different commands, which makes training more varied and enriching than a single, unchanging puzzle feeder. Additionally, the process of training is interesting and enjoyable for the owner as well as the rabbit. This will be covered further in the section on training.

Figure 3.12: Rabbits can quickly learn how to use a puzzle feeding ball filled with food.

For advice on puzzle feeding options, see Advice sheet 11.

How does diet affect rabbit behaviour?

This is the first of three comparisons of how different husbandry practices affect rabbit behaviour. In this section, we'll compare how behaviour differs in rabbits consuming different diets. In subsequent sections, we'll compare how the environment and companionship situations affect behaviour.

The diet that a rabbit eats affects its behaviour. For example, rabbits fed on rapidly consumable foods spend more time inactive. A study by Prebble et al. in 2015 looked at the behavioural differences between pairs of rabbits fed on combinations of monocomponent concentrate food (nuggets), multicomponent concentrate food (muesli) and hay. They divided the behaviours into five categories:

feeding, maintenance, inactivity, activity and investigative.

- Feeding behaviours included eating hay (from floor, from hay rack, while in hay rack), eating concentrate food, drinking water, eating bedding or eating caecotrophs.
- Maintenance behaviours included self-grooming (licking coat, washing face, scratching) and allogrooming (grooming another rabbit).
- Inactive behaviours included lying with limbs tucked under the body, lying with limbs outstretched, sitting or hiding.
- Active behaviours included hopping, chasing, jumping and stretching.
- Investigative behaviours included rearing on to the hind limbs, digging, chewing, sniffing, chin marking and paw scraping (like digging, but against the wall or door).

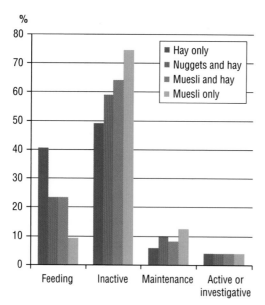

Diagram 3.3: Behavioural differences in rabbits eating different types of food (Prebble et al., 2015).

The differences in the rabbit behaviour on these diets are shown in Diagram 3.3.

We can see that rabbits fed only on hay spent much more time feeding and less time inactive. As rabbits have evolved to spend most of their time grazing, long periods of inactivity may well represent boredom. Additionally, alongside demonstrating that diet affects rabbits' behaviour, the authors also found no adverse health consequences when the rabbits were just fed on hay.

Environment

Having considered the domain of 'nutrition' in the Five Domains framework, this section focuses on the rabbit's environment. Rabbits have evolved to live in large areas of grassland.

Wild rabbits have a range of between 4000 and 20,000 sq m in grassland; this varies with food availability. They typically spend 11–13 hours per day underground in a large warren.

The optimal way to keep rabbits would be to allow them access to a large area of outdoor grassland throughout the year. The area would be secure from predators, would have areas that the rabbits could use to hide when fearful, and would allow them to run, graze and dig.

Practically, most rabbit owners cannot provide the environment described. This means that some form of compromise is necessary. The degree to which owners compromise on different environmental aspects reflects their relative importance weighting of different aspects of their rabbit's emotional and physical health. It is much easier for owners to see a risk to their rabbit's physical health from a fox than it is for them to see a risk to their rabbit's mental health from lack of access to a stimulating garden.

Common types of rabbit housing

The common types of rabbit housing are as follows:

- Outside hutch without attached exercise area
- Outside hutch with attached exercise area
- Inside accommodation without outside access
- Inside accommodation with outside access

Let's go through each of these situations in turn to assess the advantages and disadvantages of each.

Outside hutch without attached exercise area

When rabbits were kept primarily for meat, people wanted to reduce their activity as much as possible (so the rabbits converted as much of the food that they ate to increased mass) and to have them easily accessible. This led to the practice of keeping rabbits in hutches – small boxes with wire fronts for easy viewing. When rabbits started to be kept as pets rather than food animals, the hutch housing was still used. Unfortunately, keeping rabbits in this environment does not allow them to express many of their normal behaviours, so causes substantial physical and mental suffering.

The Rabbit Welfare Association & Fund (RWAF) has led an extensive campaign, entitled 'A Hutch is Not Enough', which aims to educate owners that hutches are inappropriate accommodation for rabbits. The message is becoming more widespread, but it contradicts traditional wisdom and so will take a while to be wholly accepted.

The 2016 PAW report found that 22% of rabbit owners selected an image of a small hutch (as compared to a medium or large hutch) as being most representative of the housing for their rabbit. The 2017 report found that 35% of rabbits were kept in inadequate housing.

There may be a role for a hutch as part of a larger environment (see below). However, hutches without constant access to a large run do not meet a rabbit's behavioural needs (and many that do have constant access to a run may also not meet these needs). Owners who have small hutches can be encouraged to reuse these hutches as shelters as part of a larger, more varied environment (Figure 3.14).

Outside hutch with attached exercise area

If we accept that a hutch alone is an unacceptable way to house a rabbit, then

Figure 3.13: Rabbits need to be able to stand on their back legs: the height of the enclosure must allow this.

a bare-minimum option for rabbits kept outside is a hutch and attached exercise run (Figure 3.15).

The RWAF recommends a minimum hutch size of 1.8 m long by 60 cm deep and 60 cm high (6 ft × 2 ft × 2 ft), which is permanently connected to a secure run of at least 2.4 m × 1.8 m. It suggests that this allows the rabbit some room to move (the rabbit should be able to perform three hops) and stand on their hind legs (Figure 3.13) and provides enough space to separate the food, toilet areas and sleeping areas. The RWAF points out that this size area is still very limited. In comparison to the large home range of wild rabbits, this amount of space is not sufficient to meet their behavioural requirements (their wild range is 4000–20,000 sq m). Rabbits with a permanent exercise area of this size should

Figure 3.14: Small hutches can be repurposed as shelters, but the rabbits should never be confined in such a small space.

Figure 3.15: Children's playhouses can be linked to runs using connectors to provide larger enclosures.

also be able to have supervised exercise in a larger space.

There are various ways to increase the area allocated to the rabbits. Garden sheds can be converted to many-level rabbit enclosures (Figure 3.16 and 3.17), with cat flaps out into runs (see RWAF web page on 'Converting a garden shed').

Figure 3.16: Sheds can be converted into rabbit accommodation.

Figure 3.17: Sheds don't provide a big floor area, so should be connected to a run so the rabbits have space to move around. Putting in shelves helps the rabbits to use more of the shed volume.

Inside accommodation without outside access

The 2017 PAW report found that 59% of rabbits live predominantly outside and 41% live predominantly inside the house. When owners were asked to select the image that was most similar to their rabbit's living conditions, 35% of owners indicated that the rabbit's housing was inadequate: 20% selected an image of a very small outdoor hutch, and 15% selected an image of a very small indoor cage.

Nowadays, many owners choose to keep their rabbits inside their house, so that they can spend more time with their pets. The space allocation and enclosure type for these 'house rabbits' can vary widely, but typically they will have access to a room with various types of toys and environmental enrichment, and litter trays. Owners will normally include a cage that is intended to be the rabbits' sleeping area. If the rabbits are ever to be confined within the cage, then it should be of the same minimum dimensions as described above, and should be connected to a permanent exercise area. There may be other, less visible, constraints on house rabbit environments – slippery floors, for example, decrease the effective area that the rabbits can use to display normal behaviour (McBride, 2017). If a large 'effective area' is not available, house rabbits will be as limited as those confined to a hutch outside.

Figure 3.18: Providing places for a rabbit to hide will help to reduce fear and give it more choice over its environment.

Owners may have chosen to keep their rabbits indoors for several reasons. When asked, they often mention that sharing a living space increases the time they spend with their pets, which may produce a stronger human–animal bond. As the rabbits have more space than most hutch rabbits, they tend to show more interesting normal behaviour. This is obviously an advantage for the rabbits as well, as is the increased attention that their owners pay to the cleanliness of water and bedding that they can easily see within their own house.

In the 2016 PAW report, 26% of rabbits had no interaction with humans on a daily basis, and rabbit owners were significantly more likely than cat or dog owners to say that they didn't spend enough time with their pet. Keeping the rabbits in the house will improve this.

Some owners prefer to keep rabbits inside because of risk aversion: they are concerned about the risk of predation, the risk of too low or too high temperatures and the risk of disease from other animals.

However, with protection from predators and no exposure to changing weather conditions, rabbits are at risk of boredom. As the environment will have less variation inside a house than outside a house, owners should take extra care to ensure that they provide opportunities for the rabbits to express normal behaviours. This includes providing places for the rabbits to hide (Figure 3.18), places for the rabbits to sit off the ground and plenty of places for the rabbits to forage for food.

In some ways, rabbits benefit from being 'house rabbits,' but this practice may also harm them, both physically and emotionally. House rabbits typically:

- cannot graze
- do not have access to unfiltered sunlight

- are less likely to receive enough forage
- are exposed to abnormal light–dark cycles
- show behaviours that are hard to manage inside a house
- are less likely to have rabbit companions.

Rabbits kept permanently in houses do not have access to areas of grassland for normal grazing

Grazing is important for both emotional and physical health, as described in the previous section. As well as food and space, grass is the object, or 'substrate', of many actions. Failure to provide standing grass increases the risk of redirection of normal behaviours on to abnormal substrates. This is discussed further in the section on destructive behaviours.

Rabbits kept permanently in houses do not have access to sunlight

Sunlight contains ultraviolet radiation, which is needed for the synthesis of vitamin D. Indoor lighting does not contain ultraviolet radiation (hence why you don't get sunburn from indoor lights), and ultraviolet light does not penetrate glass effectively, so sunlight coming through windows does not help. Additionally, in humans, it is known that exposure to natural light has significant effects on mood: this is through mechanisms other than the vitamin D pathway.

Research has suggested that a high percentage of house rabbits are deficient in vitamin D, and this has been linked to increased incidence of tooth disease (Fairham and Harcourt Brown, 1999). Vitamin D is important for calcium metabolism, so low vitamin D levels may also contribute to urinary tract disease (bladder sludge and bladder stones).

Humans can get some vitamin D from dietary sources, but it is very difficult to meet the daily requirement without going outside or deliberate supplementation. It is even more difficult for rabbits: the vitamin D content of a normal rabbit diet is very low (the rabbit has evolved to synthesise its own vitamin D for this reason). Additionally, it is easy to cause vitamin D toxicity in rabbits by oversupplementation (van Praag, 2014).

For good health, rabbits need to have access to unfiltered sunlight. The highest fraction of ultraviolet radiation occurs around noon: letting the rabbits outside for an hour early each morning is unlikely to have a significant effect on vitamin D levels. Owners should be advised to consider how they could give their rabbits permanent access to a safe outdoor environment. If the environment is as close as possible to their natural environment, it is much easier for them to cope, both physically and mentally, with their life as a pet.

Rabbits kept in a house may be less likely to receive sufficient forage

Owners are less likely to provide large enough quantities of hay and grass if they are concerned about creating a mess in their house: rabbits do not eat all of the forage they are given (as choice is an important part of their eating behaviour), and will distribute the remainder widely.

Rabbits kept in a house may be subjected to abnormal light–dark cycles

Some parts of the rabbit's endocrine system are affected by day length, or 'photoperiod' (Illera et al., 1992). If rabbits are subjected to differing photoperiods each day, this places stress on their hormonal systems. Owners of house rabbits should ensure that their rabbits have a reasonably constant light–dark cycle.

Some normal rabbit behaviours are challenging to manage in a house

Normal behaviours (such as urine or faecal marking) may cause the owners distress. This may either worsen the human–rabbit bond, or incentivise the owners to keep the rabbits in ways that prevent expression of these normal behaviours: convenient for the owner but challenging for the rabbit.

Rabbits kept in a house may be less likely to have rabbit companions

Some rabbit associations are still suggesting that house rabbits can be kept without rabbit companionship, which will cause substantial suffering. This arises from a misapprehension that increasing a rabbit's contact with humans can be a substitute for contact with another rabbit. This is not true, and will be discussed in depth in the section on companionship.

Often, the owners of house rabbits anthropomorphise their pets more than owners who keep rabbits in a more natural environment. Seeing their rabbits as more human can make it hard for these owners to understand that, although they are trying hard, and providing many perceived comforts for their rabbits, there are some requirements that are not being met.

If owners wish to keep rabbits as house rabbits, it is important that they have appropriate expectations of their pets.

Most house rabbits will:

- deposit faecal pellets to mark their territory, even if fully 'litter trained'
- rarely urinate outside their litter trays
- bite through thin cables
- chew and dig.

Rabbits kept in the house will deposit faecal pellets to mark their territory, even if fully 'litter trained'

These will be dry and easy to clear up. Some rabbits will do this to a lesser extent than others, but this is a normal behaviour. Owners should manage this by accepting that this is normal

and clearing up the pellets. If rabbits repeatedly leave caecotrophs on the floor (softer, mucus-covered and smellier droppings), this may be a sign of ill health.

Rabbits kept in the house will rarely urinate anywhere other than their litter trays

If the litter trays are appropriately cleaned, filled, and in an appropriate place, litter-trained rabbits should almost always urinate in the trays.

Hay on top of newspaper is the best substrate for litter trays: rabbits like to eat while they are toileting. Cat litter has been used in the past, but some cat litter pellets have been linked to respiratory problems in rabbits, and the hard edges of the pellets are less comfortable on their paws.

Owners should provide clean, hay-filled litter trays. If a rabbit urinates occasionally outside the trays, the area should be cleaned at least with vinegar (which reacts with, and removes, the calcium carbonate in the urine), and preferably with enzymatic pet urine cleaner.

Rabbits kept in the house will probably bite through thin cables

Rabbits are much more likely to bite through thin cables if they obstruct routes that the rabbits want to use.

Owners can manage this by providing plenty of hay and grass, which will reduce the behaviour if the motivation is hunger. Usually, however, the motivation is to clear routes for easier access, so rabbits will continue to remove errant 'strands' regardless of diet. Owners should ensure that cables are protected or tied out of the way.

Rabbits kept in the house will chew and dig

Like many other unwanted behaviours, chewing and digging must be understood as signals that something isn't right, rather than (as it can

Figure 3.19: Fresh branches provide both nutrition and enrichment for rabbits.

feel) a malicious attack on the owner's property! If there is not an appropriate substrate for these behaviours (appropriate from the rabbit's perspective, not the owner's), the rabbits are likely to find another outlet, such as the carpet or wallpaper.

Owners should manage this by providing the opportunity for the rabbits to dig and providing suitable substrates for chewing. Reducing concentrate foods and increasing forage feeds will reduce the desire to chew on inappropriate substrates. Fresh fruit tree branches or willow branches are very rewarding to chew and eat (Figure 3.19).

Keeping rabbits indoors can be a great experience for both pets and people, but there are also welfare costs to the rabbits and inconveniences to the owners. These inconveniences are the price that owners pay for keeping rabbits in an environment for which they are not evolved.

Inside accommodation with outside access

Another option is to combine the benefits of indoor accommodation (more human–rabbit interactions) with the benefits of outdoor exercise (access to grass and natural light), by providing litter trays, a sleeping area and food inside, with access to an outside run or secure garden. Rabbits can be easily trained to use cat flaps, which give them far more choice in their environment (Figure 3.20).

How to allow outside access

To make an environment more secure against predators, and to prevent rabbits escaping, the enclosure needs to be sturdy and well designed. This restricts the size that the enclosure can be. However, there are various methods that owners use to increase the available space for rabbits to move in.

Figure 3.20: Cat flaps allow rabbits to move between rooms or inside and outside the house as they please.

USING A HARNESS   Some owners will want to know if they should use 'rabbit harnesses' and take their rabbits for walks. Although these would seem to offer an alternative solution, they severely restrict normal behaviour and normally cause substantial distress to the rabbits. First, harnesses cause constriction around the neck and the chest of the rabbit, which is especially distressing to a prey species. Then, if the rabbit is fearful and tries to get away, the unpleasant pressure of the harness will punish the escape behaviour and worsen the fear. Finally, putting the harness on and taking it off requires restraining the rabbit and close human contact, which damages the bond between the rabbit and owner (Figure 3.21).

Anyone who has tried to train a dog to walk well on a lead will know that this can be challenging. The situation is much harder with rabbits: they do not have the same instinctive desire to follow people (so are very unlikely

Figure 3.21: Pet shops sell rabbit harnesses, but these limit a rabbit's ability to show normal behaviour and can cause significant distress when being attached and taken off.

to follow), they are usually not exposed to the lead very frequently (so will take a long time to desensitise) and have no inherent dependence on humans (so do not easily forgive stressful experiences). Harnesses are a poor solution to limited space.

Figure 3.22: There are various devices on the market to enable owners to connect hutches and runs to give their rabbits more space.

SECURELY CONNECTING DIFFERENT SPACES   A better option to give rabbits continuous access to an outdoor area is to find secure ways of linking the indoor environment with an outside run. Companies such as Runaround make systems for connecting hutches and runs (Figure 3.22). These flexible tubes and mesh tunnels can also be used to connect a cat flap (or a wall conduit, if the owner fancies a DIY project!) to an outside run to give house rabbits space outside.

Access to a garden will give rabbits significantly more space than a run, and will allow them to seek different environments when the temperature or weather changes.

RABBIT-PROOFING A GARDEN   Giving rabbits access to a whole garden is normally the best option; however, before doing this, a considered assessment of the risks should be performed. The risks involved in allowing rabbits access to a whole garden (including the chance of escape or predation) are generally outweighed by the risks involved in not doing so (including the chance of lifelong poor health and welfare). In all cases, owners should do their best to minimise likely risks, while recognising that a completely 'safe' environment may not be achievable, necessary or even desirable.

Gardens can be made more safe and secure for rabbits with a number of steps. The degree to which the garden needs to be altered depends on several aspects: how long the rabbits will spend in the garden, whether they will be supervised, the nature of the resident cat population, the types of local wildlife and the soil substrate of the garden. In order of importance, the steps to rabbit-proofing a garden are listed below. Please note that, in most cases, not all steps are required to make a garden adequately safe.

- Fix holes in fences and under gates: rabbits can squeeze through remarkably small holes. For guidance, if a rabbit can fit its

Figure 3.23: Gardens can be made rabbit-proof with sunk fences and skirts of brick or wire to discourage digging next to the fence.

head through a hole, it can probably fit its whole body through it.

- Line edge of garden (i.e. along the base of the fence or walls) with chicken wire skirt or large paving stones, or build a sunk fence to prevent rabbits from burrowing out. Be aware that chicken wire is a tangle risk and should be buried if possible. Alternatively, the wire can be attached to the lowest edge of a wooden fence and then laid on the soil, or covered with paving stones (Figure 3.23).
- Protect plants that the owner wishes to keep: helical plastic tree protectors will deter bark stripping; fencing will be necessary to protect vegetables and flowers.
- Install cat fence: fencing suitable for cats (such as the fence-top ProtectaPuss) can be used either to keep cats in one area or to keep cats out of the whole garden. In many cases, cats and rabbits will mutually avoid each other, but this cannot be assumed.
- Install bird netting: bird netting (sold to protect fruit trees from birds) can be

tied between the fence-top extensions to protect rabbits from birds of prey, if owners live in an area where this is considered an unacceptable risk.

SUPERVISING OUTDOOR ACCESS   If a garden can be made secure enough to keep rabbits in, then supervised exercise time can allow them access to far more space than could be enclosed by a run. Rabbits can be easily trained to come to a whistle for a food reward: training this behaviour makes it much easier to put the rabbits back into their enclosure when the owner has to do something else.

Regardless of their larger enclosure, rabbits must have access to a sheltered area of 'housing'. Rabbit housing has several functions. It shelters rabbits from inclement weather or extremes of temperature. It provides a place of retreat if the rabbits are fearful. It contains a predictable food source. For good welfare, rabbits should be able to choose where to spend their time: this is important for managing their temperature (moving into the sunshine or into the shade if they feel cool or warm), having access to shelter from the rain if they so wish (although many will happily choose to graze during rain showers) and being able to get away from the scent of their urine and faeces (rabbits are prey species, so prefer not to announce their presence too obviously to predators).

To improve the welfare of their rabbits, owners should consider how they can meet more of the rabbit's requirements and give the rabbits more choice over their environment. Many behavioural problems can be significantly improved by altering a rabbit's environment.

How does a rabbit's environment affect its behaviour?

This is the second of the comparisons of how different husbandry practices affect rabbit behaviour – the first section looked at diet. In this section, we'll compare how behaviour differs in rabbits in different environments. In the final section, we'll compare how the companionship situation of the rabbit affects its behaviour.

The environment of the rabbit provides opportunities to show normal behaviour. The relative effects of diet, environment and companionship on rabbit behaviour will be compared later; this section describes the effect of the environment on behaviours shown.

A study by Hansen and Berthelsen in 2000 looked at the behavioural differences between rabbits kept in a barren laboratory environment (wire cage, 46 cm × 70 cm × 40 cm; food hopper, water bottle, brick of wood and plastic flooring) and a more interesting, or 'enriched' environment (wire cage, floor area still 46 cm × 70 cm, but height 80 cm for half of the length of the cage; wooden box with plastic roof and dimensions 44 cm × 25 cm × 19 cm). All rabbits were kept singly but in close contact with other rabbits. The researchers divided the behaviours into 12 categories: feeding, maintenance, inactive, active and investigative. The differences in the behaviour of the rabbits in these two environments are shown in Diagram 3.4 (these were observed using video recordings).

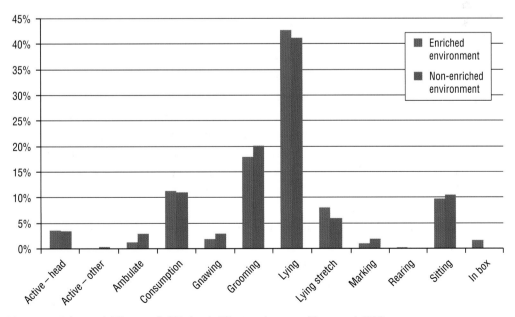

Diagram 3.4: Behavioural differences of rabbits kept in different environments (Hansen et al., 2000).

Figure 3.24: Extreme deformities, such as on this English Lop rabbit, can result in chronic discomfort from chronic ear disease and from trauma to the ear tips.

Although there were relatively few differences in behaviour between the rabbits kept in the two environments, rabbits in the enriched environment showed more 'active head' and 'rearing' behaviours than those in the barren environment. The effect of the environment seemed fairly limited (admittedly both environments were very artificial). This will be revisited later to directly compare the effect of the environment with the effect of diet and companionship on rabbit behaviour.

Health

Having considered the domains of nutrition and environment, we'll move on to exploring the domain of health. With good husbandry, pet rabbits can live for well over ten years. However, their average life expectancy in 2009 was just over four years (Schepers et al., 2009).

Wild rabbits suffer from a very different range of diseases from pet rabbits:

- Wild rabbits are at higher risk of injury or death from predation or infectious diseases (such as myxomatosis).
- Pet rabbits are at higher risk of diseases caused by:
  - poor husbandry – including poor diet (dental disease, gut stasis and urinary sludge or stones) and poor housing (brittle bones, or 'osteopaenia', and inflamed feet, or 'pododermatitis')
  - selective breeding for appearance – including ear disease in lop-eared rabbits (Figures 3.24 and 3.25), tear duct, or 'nasolacrimal duct', obstruction in short-faced, or 'brachycephalic', rabbits (Figure 3.26) and fur soiling and matting in long-haired rabbits

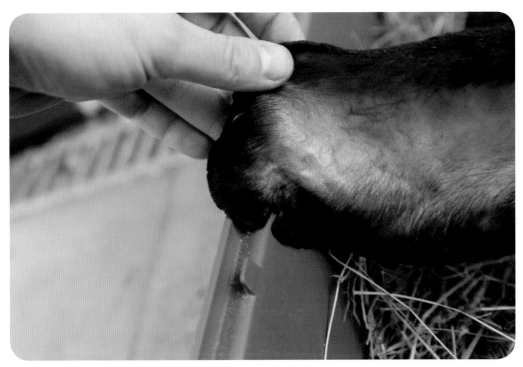

Figure 3.25: Close-up image of ear damage on the two-year-old English Lop rabbit. Although the rabbit can feel the discomfort when it stands on its own ears, the ear tips don't move when the rabbit shakes its head to try to relieve the discomfort.

Figure 3.26: Brachycephalic rabbits are at higher risk of dental disease and recurrent nasolacrimal duct blockage. Owners may also report that the rabbit 'snores' when asleep as the nasal passages are compromised.

- old age – including having longer lives to develop cancer, or 'neoplasia'.

Since many health problems are caused by poor husbandry, optimising the diet and environment should improve the physical health of a rabbit.

When assessing a rabbit during a behavioural consultation, it is just as important to check for any current physical disease (or history of disease) as it is to learn about problem behaviours (see previous section).

Taking a detailed history is very important. A history of previous husbandry-related disease (such as an incident of gut stasis, even if resolved) can be a useful indicator of current husbandry. It may also suggest the rabbit is likely to be experiencing chronic pain. These clues will also help you to identify any risks of changing the environment: a rabbit with chronic calcium deficiency due to an inappropriate diet will require a controlled husbandry change in order to prevent bone fractures: first on to the correct diet, and then, secondly, a gradual introduction to a larger area for exercise.

A full clinical exam is essential. Current disease can contribute to unwanted behaviour: pain will increase timidity or defensive behaviour (the rabbit is more vulnerable, so needs to protect itself). Also, since pain and discomfort reduce the speed of learning, current disease can reduce the likelihood of achieving an intended behaviour change.

After taking the history and performing the clinical exam, you should have identified any welfare problems arising from poor health. Sometimes, these problems may not be curable, and therefore the behaviour problem cannot be resolved. In this case, an objective assessment of the rabbit's quality of life is important.

When does poor health cause unacceptable suffering?

It may seem somewhat premature to be considering whether the rabbit will achieve acceptable quality of life at this early stage of a behavioural consultation. However, on many occasions you will already know enough to be able to understand the likely outcomes for the rabbit. In some cases, you may be asked to advise on behaviour problems caused by pre-existing health problems (for example, a rabbit showing aggression towards its owner due to its uncontrollable oral pain). In situations like this, you need to discuss the rabbit's quality of life, and whether it is ethical to continue trying to treat it.

Rabbits are motivated to perform certain behaviours – when they cannot do these, their quality of life is severely compromised. Ill health can cause unacceptable levels of suffering in three different ways.

1. When it prevents the rabbit performing rewarding behaviours
2. When it prevents the rabbit performing escape behaviours
3. When it causes chronic or unmanageable pain or discomfort.

Let's look at these reasons in a little more detail.

When it prevents the rabbit performing rewarding behaviours

Diseases that prevent a rabbit from performing rewarding behaviours will significantly affect welfare. It is easy to see that a disease that prevents a rabbit from eating causes suffering, but harder to see that a rabbit is suffering if it can no longer interact with its companion (in some cases of blindness) or if it can no longer groom itself (even if the owner still keeps it clean). It is important to understand that, just as humans cannot intuitively understand rabbits' preferred eating habits, we also may not understand how critical grooming and companion communication are to quality of life.

We need to consider how a rabbit's life will be affected if it can no longer perform a rewarding behaviour. It is important to ask questions and engage with the owner here. Quality of life is an emotive topic, but if the owner can be encouraged to remember and describe the changes, they are more likely to see the problems to the rabbit's welfare.

If a rabbit can no longer perform a behaviour because of pain or disease, then the degree to which is welfare is affected will depend on: how much the rabbit enjoyed the behaviour; how long each day it spent performing it; and how certain we can be that there aren't other welfare effects of the disease. All of these things need to be considered when making a welfare assessment. Here are some questions to ask the owner:

1. Which activity does/did the rabbit most enjoy?

   This relies on the owner's interpretation of the rabbit's behaviour, but is still useful. Asking the owner to rank the rabbit's preferred activities will help you to assess the likely impact of a disease. This is especially important if you know that the rabbit's health will deteriorate, and stop it being able to perform these behaviours. If you know which activities the rabbit really enjoys, you can use them when you're deciding on criteria for unacceptable suffering. (i.e. *He loves sitting with his companion, when he stops showing this behaviour, or if their relationship deteriorates, it is a sign that he is suffering.*)

   If a health problem affects the rabbit's most preferred activities, then the effect on the rabbit's welfare will be greater than if it affects behaviours that are not as rewarding, and the owner will understand this.

2. How long does/did the rabbit spend performing the behaviour each day?

   If a behaviour occupies large parts of the rabbit's time, even if the owner does not consider it a 'preferred activity', then restriction of this behaviour is likely to have a significant effect on the rabbit's life. A rabbit that once spent many hours each day grazing, but that can no longer perform this behaviour, is likely to suffer from boredom alongside its physical problems. A rabbit that can no longer perform normal daily behaviours (grooming itself, ingesting caecotrophs from the anus (Figure 3.27) will also suffer: evolution incentivises required behaviours by making them pleasurable.

   If a health problem affects a behaviour that the rabbit performs frequently, then the effect on the rabbit's welfare will be greater than if it affects a behaviour that is performed infrequently.

3. What changes in the rabbit's behaviour have you observed since it developed the health problem?

   This is relevant if the disease is pre-existing. With any disease, it is not possible to anticipate all of the changes that will occur to an individual animal. It may be that an animal learns to cope with a change, or it may be that the disease causes changes in ways that cannot be predicted.

   If the consequences of a health problem are unknown, it is useful to set criteria with the owner for unacceptable suffering. This gives a more objective way of assessing the welfare of an individual animal, and makes the decision for euthanasia easier if it becomes necessary.

   The answers to these three questions will help you and the owner form a picture of how much reward the rabbit has lost from its day, which will inform your understanding of its welfare.

Figure 3.27: Eating caecotrophs from the anus requires the rabbit to have good balance and flexibility. Diseases that prevent this behaviour cause ill health and poor welfare.

When it prevents the rabbit performing escape behaviours

Diseases can also cause unacceptable suffering if they prevent the rabbit from escaping from something that scares it. The rabbit is a prey species, so ability to escape from fear-inducing situations is crucial. If a rabbit is continuously exposed to a fear-inducing stimulus, without the ability to control its response, it may learn that there is nothing it can do and give up showing escape or avoidance behaviours when it is exposed to other aversive stimuli.

The rabbit will sit or lie very still. The owner may believe it is relaxed and calm. Sometimes it will be obvious that the animal is panting and holding its eyes wide open. This is called 'learned helplessness'; a close human analogy is despair or 'giving up all hope'. This occurs, for example, in rabbits that are paralysed or have had amputations that prevent their escape from handling.

Learned helplessness in animals is used as a model for depression in humans – it is the final outcome of consistent abuse. It will be very hard to resolve an unwanted behaviour in a rabbit that is suffering in this way, and the rabbit's welfare will be very poor. Rabbits need to be able to escape from perceived danger.

When it causes chronic or unmanageable pain or discomfort

Pain can be managed through healing of the disease or injury or by long-term pain-relieving medication. However, some forms of pain cannot be managed by these means, and this can be observed from the rabbit's behaviour. When animals are kept as pets, humans have an obligation to protect them from suffering. If they are experiencing pain that cannot be controlled, then this constitutes unacceptable suffering, and the rabbit should be euthanised.

Health checks

This section describes how injudicious health checks can cause more suffering than they are designed to prevent, and how they can be optimised to improve both health and welfare.

Detecting health problems promptly is important to prevent the rabbit suffering. Owners are best placed to detect changes in their pet: they are familiar with normal behaviour, and should be able to see if the behaviour changes. Owners should get into the habit of regularly assessing the health of their pet, but these health checks should be quick and easy (otherwise the owner is unlikely to perform them) and should not cause stress to the rabbit (otherwise the welfare cost of frequent health checks may outweigh the welfare benefit from improved health). A more detailed health check will be performed by the veterinary surgeon at the annual vaccination.

How should owners perform health checks?

Owners should monitor different aspects of their rabbits' health. On a daily basis, owners should monitor the feed and water intake (including the relative proportions of any dietary components), they should check the rabbits are moving around comfortably and check that there are no obvious lesions or wounds. They should also check the appearance of the faecal pellets and urine (although this may be hard to monitor). Owners can also encourage each rabbit to stand on its hind legs so any soiling or matting of the anogenital area can be assessed (Figure 3.29). The reason that these aspects of health should be checked daily is because changes to these indicate diseases in which the rabbit can deteriorate very quickly. Bodily systems can cause serious disease very quickly.

On a weekly or monthly basis, owners should check their rabbits' weight (rabbits are easily trained to hop on to scales for a food reward, Figure 3.28), should observe

Figure 3.28: Rabbits are easily trained to sit on scales for a food reward: allowing the owner to keep an eye on the rabbit's weight.

Figure 3.29: Owners can assess the underside of their rabbits by training them to stand on their hind paws for food.

Figure 3.30: Owners can palpate the teeth, jaws and skull of their rabbits if the rabbits are accustomed to pleasant interactions and trust the owners.

the underside of the rabbits' paws (when the rabbits are lying down) and should palpate the head, jaw and teeth thoroughly when grooming (to a trusting rabbit, gentle palpation around the eyes, along the upper (maxillary) and lower (mandibular) teeth and along the underside of the lower jaw is acceptable and perceived as mutual grooming, Figure 3.30). Diseases affecting these aspects of health are usually slower in progression, and so do not need to be monitored as frequently.

Older texts often advise that a health check should involve picking the rabbit up and turning it on its back to assess paw and skin health. Pragmatically, it is hard to recommend this approach: it is very unpleasant for the rabbit, and repeated attempts are likely to sensitise the rabbit to this procedure; it will become more difficult to catch, and thus the likely frequency of health checks will decrease. With good education, owners can monitor their rabbits' health very effectively in a minimally invasive way. This makes the process easier and more manageable, so owners are likely to continue doing it. This maximises the chance of the owners detecting ill health at an early stage.

Common diseases and their effect on behaviour

As described, the owner may report a previous or current disease, or you may find evidence of the disease on your clinical examination. Any concurrent disease or pain will affect a rabbit's behaviour. Below is a list of some of the common diseases of rabbits, with information on how these may affect behaviour. Since most of these are husbandry-related, there is also information on the underlying husbandry problems that lead to these. Most of these diseases have multiple contributing causes, but for simplicity, I have grouped them by major physiological or pathological cause.

Diseases caused by insufficient forage

Inappropriate diet can cause both behavioural problems directly (discussed in the previous section) and can severely affect a rabbit's health.

Dental disease

Dental disease is a severe and progressive disease of rabbits, which most commonly arises when the rabbit does not eat sufficient long fibre, usually because it is given too much concentrate food (Meredith et al., 2015). The molars grow too long, which alters the angle to which the jaw can close. This means that the incisors no longer meet and so these also overgrow, typically with the lower incisors growing outwards and the upper incisors curling into the mouth. The rabbit's jaw slowly changes shape as it attempts to eat around the longer teeth.

Vets usually treat dental disease by trimming the teeth under general anaesthesia. However, even if the teeth are trimmed back to the original length, the changed jaw will no longer open and close properly, making it very difficult for the rabbit to maintain normal tooth length. So, if a rabbit has previously had dental disease, in all likelihood, the rabbit still has dental disease.

Dental disease causes substantial pain: when the teeth do not wear down appropriately, sharp points, or 'spurs', form on the molars, which cut the cheeks and the tongue. This makes eating painful, and may cause pain even when the rabbit is not eating. When the molars grow too long, their roots put pressure on the nasolacrimal duct, blocking it and causing tears to run down the face, which damages the skin and leads to ulceration. Diseased teeth are more likely to form root abscesses, which are extremely difficult to treat in rabbits, and can lead to fractures of the jaw.

The only effective treatment for dental disease is prevention. By the time the rabbit has signs of tooth abnormalities, the damage has been done, and the mouth is likely to never return to normal. At this point, a short-term solution involves filing down the teeth under general anaesthetic, but the problem will recur. Unfortunately, these short-term treatments almost never provide the rabbit with any time when the teeth feel normal — to reduce the frequency of the operations as much as possible, the teeth are often filed down slightly more.

Owners of rabbits with dental disease typically feel very guilty, and want to continue to treat the rabbit. This requires an ever-increasing frequency of dentistry, which means that their rabbits are likely to be in continual pain (either from overgrown teeth, or from recent treatment) and should be on appropriate pain relief if their owners want to continue to manage the disease.

Obviously, dental disease causes substantial suffering. When considering the welfare of a rabbit with dental disease, you should always discuss whether the ongoing suffering from the disease and multiple surgical procedures is justified. It can be hard, or even impossible, to manage unwanted behaviours when they arise from pain.

There are various aspects of rabbit husbandry that are important for good dental health. Rabbits must have a diet consisting almost entirely of forage (which requires the normal quantity and activity of jaw movement), they should be with a companion (rabbits with companions spend a greater part of the time eating forage) and they should have access to sunlight (vitamin D deficiency has been implicated as a cause of dental disease).

Gastrointestinal stasis, ileus or intestinal obstruction

Gastrointestinal (GI) stasis, or ileus, are terms used to describe a lack of normal gut motility. Intestinal obstruction in rabbits usually occurs due to blockage with a mat of hair. Ileus can be treated medically but intestinal obstruction requires surgery. Both conditions are likely to arise when the husbandry or diet are inadequate (Meredith, 2010).

In the healthy gut, food is moved through the intestines through contractions called peristalsis. Rabbits eat large quantities of grass, which passes rapidly through the gut. A specialised area of gut sorts the partially digested food: indigestible fibre moves towards the anus for excretion as hard faecal pellets, and digestible fibre moves into the caecum, where it is fermented and the products are absorbed.

When the rabbit is resting, the gut produces another kind of pellet, caecotrophs, which contain the fermenting caecal contents (digestible fibre and microbes). These are eaten directly from the anus, and break down in the stomach, so the rabbit can absorb the protein and other nutrients.

Normally, food moves rapidly through the gut. This movement is controlled by the dietary constituents (high indigestible fibre increases gut motility) and by the autonomic nervous system (meaning that stress will reduce gut motility). If the rabbit stops eating, the motility of the gut will decrease. If the motility of the gut decreases (through stress or a low-fibre, concentrate diet), the rabbit's appetite will decrease. So GI stasis can be both a cause and a consequence of anorexia.

Although the transit of food through the gut stalls when a rabbit stops eating, the intestines will continue to extract nutrients and water from it. This continued action on the same food matter rapidly turns it into a dried-out mat that is very hard to rehydrate and break up, even if the gut does start working again. GI stasis occurs very quickly and can be fatal. A detailed discussion of treatment is beyond the scope of this book.

If a rabbit produces faecal pellets of very different sizes, or if the size suddenly changes, this may indicate gastrointestinal disease (Figure 3.31). If, when taking a behavioural history, you find that a rabbit has a history of GI stasis or intestinal disease, the first questions should be around diet. The major intervention to prevent these diseases is dietary change: aim to eliminate concentrate food and provide mostly grass or hay with some leafy greens.

Figure 3.31: Rabbit faecal pellets of different sizes may indicate problems with gastrointestinal tract health.

GI stasis can also be caused by stress. If the diet seems adequate, discuss potential sources of stress, such as pain, change of environment, changed relationship with a companion or proximity of predators. Rabbits kept singly are much more dependent on their environment for emotional stability than are rabbits kept in pairs or groups: so small environmental changes may have catastrophic effects in these animals. If a rabbit has a history of gastrointestinal disease, this may indicate an ongoing welfare problem.

Obesity
Obesity in rabbits reduces their ability to exercise, to ingest caecotrophs from the anus and to keep the anal and vulval areas clean (Figure 3.32). This in turn increases the risk of fly strike, pododermatitis, gastrointestinal stasis and ileus. One study found that 6 per cent of male pet rabbits and 11 per cent of female pet rabbits were overweight, and that

Figure 3.32: Obesity in rabbits indicates that the husbandry is deficient in some respect.

neutered rabbits were 5.4 times more likely to be overweight than entire rabbits (Courcier, E.A. et al., 2012).

Rabbits that are fed primarily on forage, are kept with companions (so have more stimulation, leading to more movement) and have constant access to a suitable exercise area are unlikely to be overweight. If a rabbit is overweight, the owner should be advised to gradually reduce and then stop feeding concentrate foods. If the owner gives a lot of 'treats', discuss training, reduce the number of treats given in a day and suggest that the owner makes the rabbit 'work' for these treats. This might be by offering the treats as rewards for wanted behaviours (such as recall), or by hiding the treats in the hay, concealing them in toys or scattering them around the enclosure to provide stimulation through puzzle feeding. These owners may be using food as a way to interact with their pet – so guiding them towards other mutually rewarding interactions can help to sustain the relationship (see Advice sheet 12 in the Appendix).

If a rabbit is overweight, there is a problem with the husbandry, which should be addressed before considering how to treat an unwanted behaviour.

Fly strike

Fly strike occurs after flies (in the UK, the blowfly *Lucilia sericata* is usually the culprit) lay eggs on fur or skin. The eggs hatch into maggots, which then attack the rabbits' skin. The flies are attracted to the smells produced by bacteria that are breaking down faeces or urine. There are various reasons that rabbits may be soiled with faeces or urine: obesity, paralysis or gastrointestinal upset (so it is difficult or impossible for the rabbit to consume caecotrophs), inappropriate husbandry (dirty environment, lack of space resulting in rabbit spending a lot of time in the litter tray) or other medical conditions (resulting in urinary or faecal incontinence).

Healthy rabbits in suitable environments should not suffer from fly strike. If there is a history of fly strike in a rabbit presenting with an unwanted behaviour, examine the husbandry. If the rabbit is overweight or has other medical problems, sort these out before addressing the unwanted behaviour.

Diseases caused by calcium imbalance

Most mammals control the amount of calcium they absorb from their gut through the action of a hormone. Rabbits, by contrast, absorb all of the calcium they ingest: this is an adaptation to their low-calcium forage diet. Whenever they absorb more than they need, they excrete the extra calcium through their urine. If rabbits are fed on a diet containing very high levels of calcium, the high levels that have to be excreted through the kidneys can cause urolithiasis (urinary stones or sludge in the bladder).

Conversely, if rabbits are fed on a diet containing very low levels of calcium (or if they do not have access to outdoor light for vitamin D synthesis, meaning they can't use their dietary calcium), they will take calcium out of their bones to meet their requirements, which causes osteopaenia (bone weakening) and fractures.

This section covers these diseases in a little more detail.

Urolithiasis

Excess calcium in a rabbit's diet can lead to calcium carbonate stones or sludge in the bladder, and this is worsened if the rabbit has a restricted water supply (a water bottle will restrict water intake compared to a water bowl), or cannot adopt a normal posture for urination (overweight or arthritic rabbits may not be able to fully empty the bladder). In severe cases, the stones can move into the urethra and cause a blockage – if this is not treated, the condition is rapidly fatal. In less severe cases the rabbit will pass 'sludge' or 'grit' in the urine, which is extremely painful but not life-threatening. There may be blood in the urine, the rabbit may strain to urinate or the owners may see 'tooth-paste-like urine'. The pain may be intermittent as the sludge moves around in the bladder (Speight, 2012).

Rabbits fed on large amounts of high-calcium foods such as alfalfa hay and some vegetables (carrot tops, spinach and brassicas including cabbage or sprouts), or those that are given a calcium or mineral supplement, are more likely to develop urolithiasis. Rabbits should eat almost entirely hay or grass; rabbits with a history of urolithiasis will probably do better on grass as it contains more water.

Current or previous urolithiasis indicates that the diet or water supply has been inadequate. This must be corrected before embarking on any behaviour modification plan. Additionally, as the condition is very painful, a detailed history of the rabbit's current urination patterns, and a thorough clinical examination, are important to ensure that the rabbit is not currently in pain.

Fractures

Fractures have a variety of causes: traumatic, developmental or pathological (fractures occurring with minimal trauma because of underlying bone weakness – osteopaenia). If a rabbit has a history of a fracture, try to find out the cause.

Traumatic fractures most commonly occur with incorrect handling: owners dropping rabbits or not restraining them correctly. If a rabbit is poorly restrained when lifted, it can kick out with sufficient force to damage its spine.

Developmental fractures occur more commonly in large-breed rabbits, where the rabbits have been selected for size and the bone strength has not increased relative to the weight. Capital femoral physeal fracture (where the head of the femur fractures) is an example of this type of fracture.

Pathological fractures occur most commonly in inactive rabbits, rabbits without access to natural light and rabbits with inappropriate calcium/phosphate intake (high concentrate diet and selective feeding). These fractures occur typically in the hind limbs or spine.

As with all husbandry-related diseases, a history of a fracture should prompt discussion of the husbandry. Some fractures are treated by amputation, but this may significantly restrict the rabbit's ability to show normal behaviour, specifically to escape from danger and to ingest caecotrophs from the anus. Owners and vets should consider the rabbit's future welfare before opting for amputation.

Infectious diseases

Infectious disease is one of the major causes of mortality in wild rabbits, and many rabbit owners are concerned about the risk that this poses to their pet. This concern about possible infection may affect how owners choose to keep their rabbits, so giving them a basic understanding of the pathophysiology of these infections is important:

Myxomatosis

Myxomatosis is caused by a myxoma virus, which leads to skin tumours, blindness, severe immunosuppression, weakness, coma and death. It is spread by direct contact with infected rabbits (the virus is shed in ocular and nasal secretions and is present in the skin lesions), by fomites (physical disease vectors such as spiny thistles or birds' feet) and, the major route of transmission, by insect vectors (such as fleas, *Cheyletiella* mites, biting flies and lice) (Meredith 2013).

In the past, myxomatosis has been used to deliberately reduce wild rabbit populations. Today there are several live vaccines available, which are given yearly, and prevent or substantially reduce clinical signs of the disease. Unvaccinated infected rabbits have a very poor prognosis, and as the disease causes substantial suffering, euthanasia is normally recommended.

Rabbits are at higher risk of infection if they are kept in an area where there are high levels of wild rabbits, especially if they have nose-to-nose contact or share the same areas of grass. Rabbits kept in more urban environments are at much lower risk, although can still be infected from insect bites or contaminated food or toys: the RWAF suggests that fleas can survive for many months in hay (Rabbit Welfare Association & Fund, 2014.).

If pet rabbits are vaccinated, the likelihood of mortality should they contract myxomatosis is much lower (in young rabbits, unvaccinated rabbits are 13 times more likely to die than vaccinated rabbits [Calvete et al., 2004]). The welfare benefits that pet rabbits get from access to grass outdoors significantly outweigh the risks of contracting myxomatosis.

Rabbit Viral Haemorrhagic Disease (RVHD)

RVHD is caused by a calicivirus and can result in death within a few days. Severe cases show fever, coma and death. Less severe cases may show anorexia, swollen eyelids, paralysis and ocular or nasal haemorrhages. The virus is spread through direct contact with an infected animal (it is secreted in respiratory secretions or discharges, urine or faeces), by airborne transmission, by contact with contaminated food or equipment or by insect vectors (as for myxomatosis). Most outbreaks occur in winter or spring.

Rabbits will be at higher risk of infection with RVHD if they come into contact with infected rabbits, and the risk factors are similar to those for myxomatosis.

As with myxomatosis, vaccination is effective, and should be repeated yearly. There are several strains of the virus: existing UK vaccines cover the RVHD1 strain, and the UK has recently approved the import of a European vaccine for RVHD2. Currently, complete protection against both strains requires different vaccines, but this is likely to change (Rabbit Welfare Association & Fund, 2016). Fear of RVHD infection should not prevent owners allowing their rabbits to have access to the outdoors.

Encephalitozoonosis

Encephalitozoonosis is a disease caused by the parasitic fungus *Encephalitozoon cuniculi*. The signs are primarily neurological (although can affect the eyes or kidneys), and typically include a head tilt and weakness or paralysis of the hind legs. Affected rabbits often become anorexic. Most rabbits are infected at an early age from the mother (the seropositivity rate in healthy pet rabbits has been reported as 30–70 per cent), but relatively few (around 5 per cent) develop clinical signs. There is no vaccine for this disease. It is not clear what causes overt disease in some rabbits, but immunocompromised animals are more at risk.

When a rabbit is stressed for a prolonged period, cortisol is released: this suppresses the immune system. If a rabbit has previously had clinical signs of encephalitozoonosis, this is a warning that the husbandry may not be suitable, and should prompt detailed questions. If a rabbit has current signs of the disease, serious consideration should be given to the welfare of the rabbit. For emotional stability, rabbits need to be able to escape from fear-inducing situations: if a rabbit is partially paralysed, it no longer has that choice. A rabbit that cannot move around cannot have good welfare, and managing unwanted behaviours in these rabbits will be very difficult.

Other diseases that can change behaviour

Arthritis

Arthritis (inflammation of the joints) is usually seen in older rabbits. It may result from periods of inappropriate husbandry and restricted exercise in early life. Arthritis causes pain. If the rabbit is in pain, it will be very difficult to change other aspects of its behaviour. When trying to manage an unwanted behaviour in an arthritic rabbit, ensure that it has appropriate pain-relieving medication.

Pododermatitis

Pododermatitis, or 'sore hocks', refers to ulcerated sores on the caudal part (heel) of the hind foot (Figure 3.33). Many pet rabbits will have a mild form: an area of hairless pink skin (Davies, O. and Dykes, L., 2000). A recent study found that 94 per cent of UK rabbits assessed had some form of lesion, from mild to severe; in neutered female rabbits, the figure was 100 per cent of those surveyed (Mancinelli et al., 2014). Open wounds are much less common. The

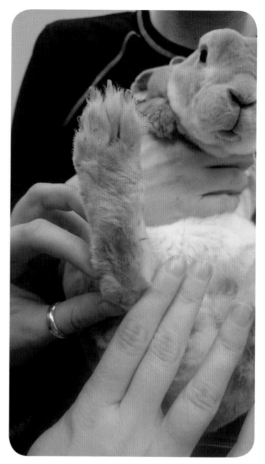

Figure 3.33: Pododermatitis in rabbits manifests as bald, reddened areas on the hind paws.

disease is currently poorly understood: it is not known whether the mild form is likely to progress to the severe form, whether the mild form causes any discomfort or whether it is possible to predict which rabbits will be severely affected.

There are many contributing factors: hard or abrasive floors increase the pressure and shear forces on the skin; long claws distribute weight on to the back of the foot; damp conditions weaken skin; increased weight increases the pressure; and immobility reduces normal skin blood flow. The disease is very hard to treat, and requires consideration of all of these factors.

Some of these factors are easy to change (claw clipping, reducing damp conditions), some require substantial environmental changes (allowing rabbits to spend more time outside) and some factors cannot be changed (there is only so much weight that a large-breed rabbit can lose).

In the rabbit's normal environment (grassland), it will cover a large distance every day when grazing, and will choose to rest in areas that are dry underfoot. The claws will normally sink into the grass to provide anchoring points from which to propel the rabbit forwards.

Since almost all rabbits in captivity have signs of pododermatitis, and as it is not clear whether the condition is progressive, frequently turning rabbits upside down by owners to check paws (as is recommended in many textbooks) can't be justified – it should be assumed that all rabbits are likely to have some degree of pododermatitis, and so all rabbit environments should be improved to reduce the risk. Owners should be advised to monitor the rabbit's behaviour carefully and check the paws thoroughly if they see any signs of change in gait, or reluctance to move. Additionally, the paws should be checked during the clinical examination for a behavioural consultation: pododermatitis can cause chronic pain.

Physical/functional domain: Situation-related factors

Pain and disease affect the rabbit's internal physical or functional state. The other form of sensory input is external – it arises from how the rabbit interacts with its environment and whether it is able to express rewarding behaviours. A previous section briefly discussed this in the context of the pleasures arising from different aspects of the nutrition, environment and health, but this section will explore how the rabbit interacts with different parts of its situation: the physical environment, other rabbits, humans and other pet species. It will explain how different interactions may cause positive or negative affects.

Interactions with physical environment

The environment in which a rabbit lives is very important to its physical and mental health. Wild rabbits spend a lot of time establishing and maintaining territories, which are defended against unfamiliar rabbits. These territories are about 4000 sq m (Southern, 1940), and provide the rabbits both with food and with shelter from inclement conditions and predators. The pet rabbit's environment is usually many times smaller than that of a wild rabbit, and usually contains far less stimulation.

The environment of pet rabbits is limited for various reasons.

- Lack of knowledge (the owner doesn't realise the amount of space that rabbits need).
- Lack of space (the owner has no more space to offer).
- Lack of money for financial outlay (the owner doesn't feel that they can afford to prepare a larger area).
- Concerns over safety (as described earlier, the owner feels that an increased risk of predation outweighs the benefits of a larger environment).

Rewarding rabbit–environment interactions
As rabbits can't verbally explain whether or not they enjoy a certain interaction, humans have to make assumptions about their emotional

state based on their behaviour. Initial welfare assessments focused on the absence of behaviours that indicated negative emotion (i.e. no thumping, which would indicate that the rabbit was feeling stressed). Current methods assess the experience of the rabbit through a number of behavioural expressions of positive emotions.

Boissy et al. (2007) suggested that key behavioural indicators of pleasure are play behaviours, affiliative behaviours (covered in the section on companionship), self-grooming, vocalisation and information gathering.

Play behaviours

Play behaviours include 'functional' behaviours (such as fleeing, fighting or sexual behaviours) as well as specific play behaviours. As described later, young rabbits show specific play behaviours in the form of chasing and displacing other rabbits, but adult rabbits show few specific play behaviours. However, functional behaviours in adults may be exaggerated, repeated or more variable, and may occur without the 'normal' stimulus or consequence. Binkying (a twisting or jumping action when the rabbit is relaxed, without an obvious trigger, see Figure 3.34) is a prime example of this form of play behaviour. Functional play behaviours are thought to train the skeletal muscles, reflexes and emotional responses for unexpected events.

Conditions associated with poor welfare, such as very restricted hutches, suppress play behaviours in rabbits. If rabbits are housed in a way that restricts play, there will be a rebound spike in play behaviours when they have access to a more rewarding environment. Owners may see this when the rabbit binkies round the garden when it is released from its hutch. Play behaviours are rewarding to rabbits, and providing an environment that allows

Figure 3.34: A binky is a twisting or jumping action performed without an obvious trigger. Photo courtesy of Lena Bergmeister.

expression of these behaviours will improve their welfare.

Most rabbits like to dig, and so expression of this behaviour should be encouraged in a well-planned rabbit environment. Digging is a functional behaviour, but many rabbits will also dig scrapes to roll in or fully lie down in: both relaxed behaviours that indicate pleasure.

Most of the functional play behaviours of the rabbit include running or jumping: short sprints, running with a companion or binkying. These require sufficient space: the minimum run size specified by the RWAF is unlikely to be enough. The larger the area that the rabbits can access (supervised, or even better, unsupervised), the more they will be able to express these behaviours.

Self grooming

Self-grooming behaviours refers to the behaviours that an animal performs to maintain its own body surface: in rabbits, this means licking, scratching and rubbing of the fur. Self-grooming behaviours are rewarding to the rabbit, but high levels of self-grooming behaviour may also be performed as a displacement activity (levels are higher in rabbits kept in barren or restrictive environments, and in rabbits deprived of access to hay, Figure 3.35). So judging the quality of an environment based on the rabbit's self-grooming behaviour can be difficult.

However, self-grooming behaviours are stimulated by water or dirt on the fur, and grooming to remove water and dirt is very functional and unlikely to be a displacement activity. Providing an environment in which the rabbits can get muddy or wet (should they choose to do so) and then clean themselves allows normal behavioural expression and increases the reward of self-grooming.

Figure 3.35: Self-grooming can indicate good welfare or poor welfare, so should not be used as an indicator in isolation.

Vocalisation

Rabbits rarely vocalise. Of their (infrequent) vocalisations, some are associated with negative experiences (growling or screaming), some are associated with sexual behaviour (grunting) and gentle chattering of the teeth occurs in pleasurable situations: when resting, after completing self-grooming or when being groomed by another. So the latter can be considered as an indicator of pleasure.

Information-gathering behaviour

Such behaviour, or exploration, is a behavioural need in almost all animals, and it is closely related to and affected by fear. There are two types of exploration – inquisitive and inspective. Inquisitive exploration is performed when the rabbit is looking for a change, and inspective exploration occurs when the rabbit responds to a change. Rabbits explore when they have no other more immediate needs, and may show this behaviour for a prolonged period, which suggests that it is pleasurable.

When designing an environment that allows exploratory behaviour, the owner needs to strike a balance between allowing the rabbits to feel curiosity, and causing fear through unpredictability and continuous change. Ideally, the major features of the environment will remain predictable, but minor aspects should vary frequently to prevent boredom.

It is easier to provide ongoing mild variation for outdoor rabbits than indoor rabbits – the changes of weather and season will do the job. However, the environment inside a house is maintained within very narrow environmental parameters: the temperature is stable, the light levels are often stable and the rabbits never encounter rain or wind.

This reduces the drive for exploratory behaviour and reduces the need for rabbits to use behavioural methods to maintain homeostasis (sheltering from rain, crowding together, sunbathing, facing upwind). It is intuitively obvious that an environment that is exposed to the elements without shelter will compromise welfare, but it is less obvious that a completely controlled environment will also reduce rabbits' ability to perform normal behaviours. A happy medium is an environment that contains suitable areas for the rabbit to shelter if necessary, but that also provides areas that are exposed to the elements (Figure 3.36).

Providing this sort of environment allows the owners to assess a rabbit's preferences. Many owners are surprised that their rabbits will choose to graze outside in bad weather, seeking shelter only when the rain starts to seep through the coat, or when the rain becomes uncomfortably heavy. However, this gives important insight into the rabbit's choices – something that is not possible in a completely controlled environment.

An ideal environment should allow rabbits to exercise a degree of choice over their response to different environmental stimuli, to experience reward when they seek shelter and to explore a predictable yet constantly changing environment.

Aversive rabbit–environment experiences

Poorly designed environments can cause substantial distress to a rabbit. The environment can cause fear, frustration and pain.

Environment causes fear

This occurs when the rabbits are housed within sight, sound or smell of predator species that show predatory behaviour towards them. The stress is compounded if the rabbits do not have space to move away from the predator or areas in which they can hide from it. A study looking at the 'costs' that a rabbit is prepared to pay to have access to a raised platform found

Figure 3.36: Seasonal and daily variation in weather and climatic conditions provides enrichment. Owners who give their rabbits choice may be surprised that their rabbit chooses to spend time outside in adverse weather: but this choice is important for their welfare.

that this is almost as motivating as access to food (Seaman et al., 2008), even if the rabbit rarely uses it. Rabbits need to be able to show vigilance and sentinel behaviour (sitting and watching from raised areas, Figure 3.37) and need to be able to flee for cover (pipes, boxes or undergrowth).

Environment causes frustration

Various environments can cause frustration: very small environments (many commercially available hutches and runs are woefully small, Figure 3.38), very barren environments and lack of appropriate bedding.

When considering what constitutes a barren environment to a rabbit, consider the environment from the rabbit's perspective, not the human's perspective. An enclosed area of lawn will be much more stimulating for rabbits than an equally sized indoor room, even if both lack objects to explore. Owners should consider how to make the best use of the space and objects that they have (Figure 3.39).

Environment causes pain

Unsuitable environments can cause injury to rabbits from sharp edges on wire or abrasive flooring. Environments that are not cleaned sufficiently increase the risk of disease. Rabbits try to move away from latrine sites, as the smell can attract predators – so keeping rabbits in an dirty environment may also cause stress when they are confined in a strong-smelling area.

Providing sufficient space will lessen the effect of a poor environment, as the rabbits then have some degree of choice to avoid the unpleasant situation.

Figure 3.37: Not all rabbits will choose to rest on a raised platform, but they will work to have access to one.

Figure 3.38: Many commercially available runs are woefully small.

Figure 3.39: Rabbits should have access to objects that allow them to express normal behaviours. Pipes (appropriate to the size of the rabbit) make excellent tunnels for enrichment and shelter.

Interactions with other rabbits

In this section, we'll understand why companionship is so important for rabbits, and why keeping them on their own causes such major welfare problems.

Rabbits have evolved to live in social groups. Wild rabbits live in stable breeding groups consisting of 1–8 males and 1–12 females. The average group sex ratio is 1.5 females per male.

A combination of better owner education and changes in marketing of rabbits (for example, substantial discounts on pairs of rabbits bought compared to single rabbits) has had an effect on the percentage of rabbits kept singly. However, this is still currently the most common situation in which rabbits are kept (52 per cent of UK rabbits, according to the 2016 PDSA Animal Wellbeing report).

Rabbits are very social animals. In the wild, they live in a large warren but have a close group of five or six animals within the colony. When rabbits were domesticated as food animals, it became traditional to keep rabbits in solitary confinement in a small cage. Unfortunately, many people are still keeping rabbits in this way (Figure 3.40). An increasing body of evidence about the welfare and behavioural needs of animals demonstrates that keeping rabbits on their own causes a lot of preventable suffering.

Figure 3.40: Some hutches for sale are far too small to be used for rabbits. However, uninformed owners will not be aware of the welfare problems caused by this sort of housing.

Rabbits kept singly

Rabbits kept singly have shorter life expectancies

A study of almost a thousand pet rabbits in the Netherlands (Schepers et al., 2009) found that about half of the respondents kept their rabbits singly, and most of those were kept in small cages. This study found that the mean age at which pet rabbits die is approximately 4.2 years, but this was 3.3 years for rabbits kept singly and 5.1 years for rabbits kept with other rabbits.

Rabbits kept singly try really hard to see other rabbits

Another study looked at the 'costs' that rabbits were prepared to pay to get access to different desired resources, including food and limited social contact with another rabbit (Seaman et al., 2008). The 'cost' that needed to be paid was the difficulty of pushing through a weighted door. Rabbits worked almost as hard for access to social contact as they did to get food. As mentioned in the previous section, affiliative behaviours (staying close to each other or allogrooming between specific individuals) indicate a positive emotional state.

Rabbits kept singly show more abnormal behaviours

As singly housed rabbits spend less time grazing, and cannot express normal social behaviour, they have more time that is spent inactive and alert. This inability to express normal behaviours is likely to contribute to boredom and lack of normal stimulation. Pet rabbits that are kept singly sat up more when placed in an open area, indicating that they are generally more fearful (Schepers et al., 2009). Singly housed rabbits are more likely to be 'destructive', as they spend proportionally more time interacting with their environment to substitute for the time spent in social contact. One study (Mullan and Main, 2007) found that singly housed rabbits were also more likely to tolerate being picked up by their owners: perhaps because of the frustrated need for social contact in a socially dependent species. However, another study (d'Ovidio et al., 2016) found that owner-directed aggression was significantly more frequent in singly housed does than those kept with a rabbit companion. Keeping rabbits singly may enhance their fear of humans, leading them to develop aggressive coping strategies.

Rabbits kept in groups

Rabbits kept in groups exercise more

One study looked at differences in the physiological and behavioural responses of young female rabbits housed singly or housed in groups (Whary et al., 1993). These rabbits tended to socialise in small groups (one to three rabbits). The researchers did not see overt aggression or competition for space, feed or water access during the study period. Group-housed rabbits ate significantly more, but the growth rate was the same as singly housed rabbits, probably because of increased exercise.

Rabbits kept in groups graze more

Rabbits graze almost continuously while awake, which means that they need to be above ground and are therefore vulnerable to predators. They therefore rely on being in a group, where many eyes and many ears can be alert for predators. When one rabbit spots a predator, it thumps to provide an audible alert and then runs for cover. In doing so, their movement and the white underside to their tail (the scut) provide a visible alert.

It is reasonable to think that knowing that there are other rabbits around, which would alert them to the appearance of a predator, allows a grazing rabbit to relax slightly. Certainly, when rabbits are kept alone, they spend a greater amount of their time alert and watching for danger. This reduces the time they can spend eating, which means that they are more likely to seek high-calorie food that requires less time to chew: food that is more likely to cause dental disease, obesity and ileus (gut stasis).

Rabbits kept in groups cope better with cold weather

Rabbits maintain body temperature in a variety of different ways. Their thick fur provides insulation, but they also show behavioural thermoregulation. In the wild, rabbits can retreat into underground warrens where the temperature is more stable. This is not available for the majority of pet rabbits kept outside. Wild rabbits also huddle against each other to conserve heat, reducing the surface area for heat loss and sharing body heat to a certain extent. Pet rabbits kept outside alone are additionally deprived of this method of maintaining body heat, so they are more vulnerable to the cold.

Rabbits kept in groups have better eye health

Rabbits frequently groom each other around the eyes. As well as strengthening social bonds, this behaviour cleans the eyes and removes foreign bodies. This may account for anecdotal reports of worsening eye health in rabbits when their companion dies. Since rabbits are heavily reliant on their vision to spot danger, and therefore good eye health is vital, nature has made eye grooming very pleasurable for rabbits. Indeed, when compared to dogs and cats, rabbits tend to object a lot less to palpation of the eyeball and socket.

There is a large body of evidence that demonstrates rabbits suffer from social isolation.

Depriving a rabbit of social companionship is hugely detrimental for its welfare, and should never be advised. Any owner who keeps a rabbit singly should be strongly advised to acquire a suitable rabbit companion, for both their rabbit's welfare, and to improve their own rabbit-owning experience.

Case study 2: Bailey

Bailey was a male lop-eared rabbit, who was kept singly for two years with a lot of social contact with humans. He was nervous and highly reactive to novel sights and sounds. When rehomed at the age of three, he initially would bite, growl and scratch when handled. His environment was changed so he no longer needed to be picked up, and he was clicker trained. The aggressive behaviours stopped, but he remained flighty and reactive.

After six months, he was neutered. He became even less confident and was more reluctant to approach people. Two months after being neutered, he was bonded to a neutered female rabbit at a rescue centre. The bonding process was uneventful. His behaviour changed markedly within the first few weeks: he began to lie down fully (something that he had rarely done before) and he became significantly more relaxed around familiar and unfamiliar people.

Relationships between rabbits

In a social relationship between two animals, one animal usually has preferential access to resources over another. This animal is said to have a higher social status, or be 'dominant' over the other, and the other animal reacts submissively to the first – it is 'subordinate'. If both animals had to fight for resources every time they were available, they would waste energy and risk being injured. Having a defined hierarchy between a pair of animals increases the fitness of both animals by reducing conflict. The terms 'dominant' and 'subordinate' describe the relationship, not the animal. A rabbit may be dominant in a relationship with a specific rabbit, but it may well be subordinate to another rabbit.

When rabbits are introduced, part of the bonding process involves establishment of the relationship hierarchy, usually by threatening behaviours (such as chasing) or interchanges (both rabbits with lowered heads, waiting to be groomed). Once this status is established, the relationship is maintained by agonistic (competitive) behaviours, with aggressive behaviours much reduced or even absent. The dominant rabbit may occasionally displace the subordinate rabbit from a food resource, or may chase it. The subordinate rabbit usually defers or withdraws if the dominant rabbit shows agonistic behaviour, which signals understanding of the hierarchy and reduces the frequency of these displays.

Owners can determine which of their rabbits is dominant and which subordinate. The dominant rabbit in a relationship is usually the one that is more likely to be groomed than to groom when the rabbits interact, and it is more likely to feed first if offered a bowl. In some cases, it is more likely to lie with limbs outstretched when the two rabbits are together, which increases the requirement for the subordinate rabbit to remain alert for danger. The dominant rabbit in a relationship may be more likely to displace the subordinate rabbit when interacting with humans (Figure 3.41).

Owners should respect the hierarchy between their rabbits and not attempt to

Figure 3.41: The dominant rabbit in a relationship is more likely to be groomed than it is to groom, is more likely to eat first and is more likely to displace the subordinate rabbit from interactions with the owner.

disrupt it to 'be fair' to the rabbits: they have formed their relationship, and attempts to alter it may disrupt the bond between the rabbits, and result in aggressive behaviour from the dominant to the subordinate rabbit. The owners can ensure that the subordinate rabbit gets a suitable share of food by giving food items to the dominant rabbit first and then immediately to the subordinate. Also, the owner can reinforce the bond between the rabbits by grooming both rabbits when they are sitting next to each other (Figure 3.42).

The social hierarchy is context dependent in various respects. It is specific to a certain relationship: a rabbit that is dominant in one relationship may well be subordinate in another pairing. It can also vary depending on the situation, location and type of resource. Subordinate rabbits may also show differing responses depending on the situation and value of the resource: if the dominant rabbit shows agonistic behaviour over a less-favoured resource (a handful of picked grass), the subordinate rabbit may show passive submissive behaviour and withdraw, but if the resource is of higher value (a piece of apple peel), the rabbit may show active subdominant behaviour, snatching the food and running away with it.

These interactions are necessary for a healthy relationship between rabbits, and form the basis for pleasurable interactions such as allogrooming (social grooming between members of the same species).

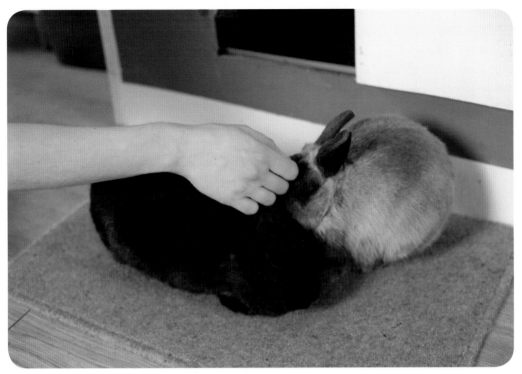

Figure 3.42: The owner can reinforce the bond between the rabbits by grooming them both when they are resting together.

How does companionship affect rabbit behaviour?

This is the final comparison of how different husbandry practices affect rabbit behaviour – the first sections looked at diet and environment. In this section, we'll compare how behaviour differs in rabbits kept singly or with other rabbits.

One study (Podberscek et al., 1991) looked at the behavioural differences between rabbits kept on their own and rabbits kept in groups. Rabbits kept with other rabbits spent significantly longer performing comfort behaviours (such as self-grooming or allogrooming) and marking or investigatory behaviours, and less time performing maintenance behaviours (such as standing, resting, feeding and eliminating) (see Diagram 3.5). Agonistic behaviours are only performed towards another rabbit, so by definition only occurred in the rabbits kept in a group. Stereotypic behaviours were not observed at all in rabbits kept in groups, but rabbits kept on their own spent 6 per cent of their time performing these behaviours. This has implications for management of stereotypies in pet rabbits.

Although wild rabbits may have four or five close companions, it is rarely practical to keep this many rabbits in a group as pets, due to space constraints – very large areas are required to allow the required retreat behaviours during large group hierarchy formation. Therefore, pet rabbits should normally be kept in pairs, to give the greatest chance of a stable relationship.

Why do owners keep rabbits singly?

Despite the overwhelming evidence that rabbits require the companionship of other rabbits for good welfare, there are many owners who still choose to keep them in social isolation. There

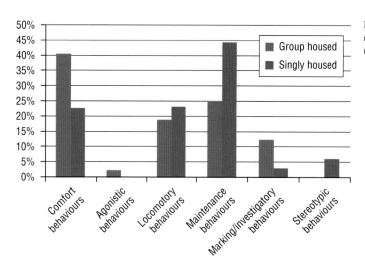

Diagram 3.5: Behavioural differences in rabbits kept singly or in groups of rabbits (Podberscek et al., 1991).

are various reasons that an owner might keep a rabbit alone.

- Owner lacks knowledge about the social needs of their rabbit.
- Owner believes that they can personally meet the social needs of their rabbit.
- Owner doesn't want to own rabbits forever (so doesn't want to replace an animal that has died).

Let's explore each of these cases in turn.

> *The 2016 PAW report found that 52% of rabbits were still kept alone. The 2017 report found that owners of male rabbits were significantly more likely to house them alone (67%) than the owners of female rabbits (35%).*

Owner lacks knowledge about the social needs of their rabbit

These owners do not know that rabbits should be kept in pairs. These owners are likely to have bought a rabbit as a pet for a child, and are unlikely to have done significant pre-purchase

research on the species. They may rely on knowledge from popular culture or on the experiences that they had with rabbits as a child. When they are advised that they should keep two rabbits rather than one rabbit, they are likely to cite problems with increased space requirements (*'our hutch is too small for two rabbits'*), increased cost of care, and previous experience with rabbits (*We had a rabbit as a child, he always lived on his own and he was eight when he died*).

Their reasons are often based on their limited expectations of the rabbit as a pet, and so their assessment of welfare is also limited. The challenge here is that owners that have low expectations of the rabbit are often not willing to spend time or money to improve the rabbit's quality of life. They may be able to acknowledge that the situation is not perfect from the rabbit's perspective, but feel that they lack the capability to improve the situation. If an owner has very low expectations of their pet, they are unlikely to have a strong bond with it. If so, appealing to their relationship with the rabbit will be ineffective, as this relationship is not a strong motivator. A more successful approach might be to highlight the improvement in the rabbit's quality of life without additional demands on the owner's time.

Owner believes that they can personally meet the social needs of their rabbit

These owners are better informed about the needs of a rabbit, and know that rabbits are a very social species, but believe that they can fulfil this requirement of their pet by their own companionship. Many older texts and some websites still recommend that rabbits are kept in pairs when kept outside (due to the higher risk of neglect) but say that it is acceptable to keep single rabbits as house rabbits. This advice is no longer seen as acceptable by rabbit welfare associations, such as the RWAF, but as recommendations have changed substantially over the last few years, many owners may not be aware of this.

Such owners are often very invested in their pet, and may fear that having two rabbits will reduce the attention and affection that their rabbit feels for them personally. They commonly give reasons such as '*He's very dominant, he hates sharing his space, and I don't think*

he'd like another rabbit', or '*She's such a happy rabbit, she plays all of the time.*'

They are less likely to frame reasons in terms of difficulties with an additional rabbit, and more likely to see the current situation as being the perfect life for their rabbit. These owners often have very high expectations of their pet, and see their rabbit as being 'just like a dog'.

Such owners will often spend large amounts of money to alleviate boredom in their rabbits through environmental enrichment, but will often not see that the best way that they can provide for their rabbit's needs is by acquiring another rabbit. These owners are often worried that a new rabbit may 'come between' them and their pet. Therefore, describing the 'time budget' or 'energy budget' of a rabbit may be a useful way of framing this argument – increasing the 'budget' available for interaction with the owner by reducing the stress of the rabbit is likely to improve the relationship with the owner (Diagram 3.6).

The challenge here is that owners are unlikely to see that a problem exists, having never known a rabbit that was kept with another

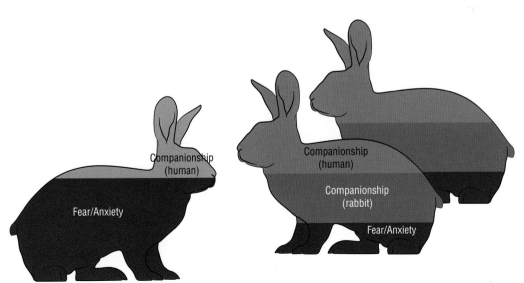

Diagram 3.6: Time budget of singly housed and group-housed rabbits. Singly housed rabbits spend proportionally much more time alert and fearful, reducing the time budget that they have for normal relaxed interactions with the owner.

rabbit. They may not want to acknowledge there is a problem as they are heavily invested in their rabbit, and if they acknowledge that they can improve the situation for their animal, they are inviting retrospective guilt. Therefore, any solutions should avoid criticism of the current situation and should instead align with the owner's motivations. Owners often feel guilty when they leave their pets when they go on holiday – a companion rabbit will reduce the pet's loneliness when the owner is away. These owners are often very proud of the life that they provide for their rabbit – encouraging them to rehome another rabbit to improve its quality of life may also be successful. Providing companionship for their existing rabbit will then be a secondary outcome.

Owner doesn't want to own rabbits forever (so doesn't want to replace an animal that has died)

Another group of owners may have previously had a pair of rabbits, but lost one. These owners may fear the time cost of bonding another rabbit, they may think that their current rabbit may soon die anyway, and they may be worried about the perpetual cycle of owning rabbits – they may not see a way out! For these owners, discussion of rescue centres or rehoming the rabbit, as compared to taking on another rabbit, may help them to understand the welfare cost of isolation while empathising with the situation in which they find themselves (after all, they have already once done the right thing). Some rescue centres may foster out older rabbits, which will be taken back when the owner's pet dies, thus relieving them of the burden of caring for another rabbit for the rest of its life. The challenge here is demonstrating to the owners why they should consider keeping two rabbits together, and if they don't want to, discussing the options for rehoming their rabbit to give it the chance of a life with a companion.

In summary, the approaches to encouraging owners to keep two rabbits are outlined in Table 3.2.

It's important to discuss the motivations of the owner and how those relate to how they

Table 3.2 Approaches for encouraging owners to keep rabbits in pairs.

Cause	Challenge	Approach
Lack of knowledge	Low willingness to invest time or effort in pet	Emphasise improvement in quality of life for pet without requiring additional engagement from owner
Belief that owner can meet rabbit's social needs	Fear that having two rabbits will impair the owner's relationship with the rabbit	Highlight welfare gains for pet while owner is on holiday
		Appeal to owner's desire to improve rabbit welfare by suggesting that they rehome another rabbit
Owner does not want to replace a deceased rabbit from a pair	Owner fears perpetually owning rabbits	Suggest that the owner rehomes or fosters an older rabbit
		Suggest that the owner relinquishes their remaining rabbit for rehoming

choose to keep their pet. Showing empathy here will help you to understand their concerns and their desires, so you can tailor your advice based on this.

Mutually rewarding rabbit–rabbit interactions

It can be difficult for owners to appreciate what sort of rabbit–rabbit interactions are likely to be pleasurable, as rabbit motivations are very different from human ones. However, it can be assumed that common, non-agonistic social interactions are pleasurable – if a behaviour is perceived as pleasurable, it is more likely to be performed.

Pleasurable interactions are:

- allogrooming
- resting in proximity
- visual contact with another rabbit.

Allogrooming

Allogrooming refers to social grooming between members of the same species. In rabbits, this behaviour is focused on the face and ears.

Sometimes, grooming may be spontaneously offered, typically when two rabbits are in close proximity, one is washing itself, and grooms the other rabbit in passing. Sometimes one rabbit will actively solicit grooming by lowering the head and ears towards the other – if the second rabbit participates, it will usually lick the first rabbit on the top of the head, around the ears or around the eyes (Figure 3.43 and Figure 3.44). It may also respond by putting its own head down – usually the rabbit that can get the head lowest is successful in being groomed. Within a relationship, the rabbit that is more dominant in that relationship is more likely to be groomed by the other rabbit, but owners are likely to observe grooming in both directions within a pair of rabbits.

Some rabbits, especially those that were socially isolated at a young age and then reintroduced to other rabbits later in life, may direct this behaviour at other parts of the anatomy, such as the flanks or tail. This is rarely tolerated by the other rabbit being groomed, which will usually move away. Attention on the back and rump is usually sexually motivated.

Figure 3.43: Rabbits actively solicit grooming by lowering their head and ears towards the other rabbit.

Figure 3.44: Rabbits tend to groom each other on the top of the head, along the eyelids and on the ears.

Figure 3.45: Resting in proximity to a familiar rabbit is a rewarding behaviour.

Most owners know that this behaviour is pleasurable for the rabbit being groomed: it usually occurs when the animals are relaxed, is often actively solicited and the groomed rabbit may show signs of pleasure (such as gentle grinding of the teeth).

Resting in proximity

Resting in proximity to a familiar rabbit is a rewarding behaviour. When given a large space, a pair of rabbits may not always spend time in close proximity, and may have different preferred places to rest. However, rabbits do show behavioural thermoregulation by huddling, and it has been suggested that singly housed rabbits are more susceptible to cold temperatures than are rabbits housed in pairs or groups, which continue to rest in close proximity to share body heat.

In addition to thermoregulatory benefits, rabbits that rest together have a higher chance of perceiving danger, which increases the time that they can spend relaxed. Many owners report that, when they get a companion for a previously singly housed rabbit, the rabbit spends more time in relaxed positions (Figure 3.45).

Visual contact with another rabbit

It can be hard for owners to see that it is mutually rewarding for a rabbit to be in visual contact with another rabbit, but this makes evolutionary sense. Research has suggested that rabbits will work almost as hard for visual, olfactory and tactile contact with another rabbit as they will for food, even if they don't know the other rabbit (Seaman et al., 2008). When a rabbit can see another rabbit, its stress levels will be lower, so it is more likely to show normal investigatory behaviours, and to follow the lead of the other rabbit (Figure 3.46).

Aversive rabbit–rabbit interactions

Aversive rabbit–rabbit interactions include:

* agonistic behaviour from one rabbit
* territorial behaviour from one rabbit
* mounting
* fighting.

Agonistic behaviour from one rabbit

In a bonded pair of rabbits, it's normal to see some agonistic behaviours – behaviours relating to fighting (not just aggressive behaviours, also threats, retreats, placation and conciliation). These are usually performed by the rabbit that is dominant in the relationship in response to a disliked behaviour of its companion.

Agonistic behaviour may be shown during competition for a high-value food or access to a favoured location, object or person. Typical agonistic behaviours include ears flattened to the head, sharp movements or lunges and, in some rabbits, a growl. The subordinate rabbit usually defers to the dominant rabbit by giving up access to the resource. Agonistic behaviour from one rabbit is only a problem if the other

Figure 3.46: Rabbits show more exploratory behaviours when kept with other rabbits.

rabbit also responds agonistically, as a fight may result.

This is a normal and healthy part of a rabbit relationship. Owners should be discouraged from intervening or punishing this behaviour, as a stable social hierarchy is very important for emotional stability. Owners can lessen the incidence of this behaviour by giving food rewards to both rabbits at the same time (the rabbits will typically eat small pieces of food at the hand, or will take larger pieces and move some distance apart) or by scattering the food on the ground to encourage foraging behaviour rather than direct competition.

Territorial behaviour from one rabbit

If an unfamiliar rabbit is introduced on to another rabbit's territory, the resident rabbit will often approach the new rabbit, sometimes nibbling at grass, marking different objects or digging. Territorial behaviours include chasing, scraping of the ground and stiff-legged runs past the new rabbit. If the new rabbit does not retreat (as it would usually do), the rabbits will fight.

Mounting

Mounting, or 'humping', can be a normal behaviour between pairs of bonded, neutered rabbits. A rabbit of either sex may mount their companion, either from the front or from the back. This behaviour is much more common during the bonding process, and may take several weeks to reduce. Mounting behaviour also occurs more commonly in spring. As with agonistic behaviour, mounting is only a problem if the other rabbit responds aggressively, as the mounting rabbit may be injured.

Fighting

True rabbit fights can be very severe and may even be fatal. When fighting, rabbits bite each other's necks with their teeth while kicking powerfully with the hind legs. Fights may occur when a new rabbit is introduced on to another rabbit's territory or in a bonded pair if neither rabbit displays submissive behaviour. Behavioural changes (such as from neurological problems or pain) or hormonal changes (that occur around puberty) may result in one rabbit perceiving the other rabbit as unfamiliar, which may also trigger a fight.

Owners may describe various kinds of agonistic behaviour as a 'fight', and it is important to ask specifically what is meant by this term. True fights involve definite attempts to injure the other rabbit, often around the face, neck and flanks. Fur is often pulled out from these areas. During the bonding process, some male rabbits will mount the female and pull fur from the middle of the back while mounting: the presence of fur on the ground does not guarantee that a true fight has occurred.

Rabbits will fight with unfamiliar rabbits on their territory, which is why bonding a pair of rabbits is best undertaken in an environment that is novel to both animals. Fighting should never occur in a bonded pair of rabbits: fighting indicates that the normal social communications that prevent fighting have broken down. Owners should not be advised to rehome a pair of prepubescent male rabbits, as the hormonal changes at puberty may trigger a fight.

Interactions with humans

Owners acquire pet rabbits (usually) because they want to interact with the rabbits. These interactions are very important: they strengthen the human–animal bond, which increases the owner's motivation to care for the rabbit. When appropriate, these interactions provide enrichment and reward for the rabbit.

Case study 3: Emmett and Merrill

Emmett and Merrill were a pair of three-year-old Dutch rabbits that had been rehomed about a year previously. The owner reported that the rabbits had been fearful of everything when they were brought home, but had subsequently relaxed. However, she was concerned that they were completely uninterested in people. They would take food from the hand, but when the owner tried to stroke or touch them, the rabbits would jump away, thump and hide. The rabbits lived in the house and the owners worked from home, so they had frequent exposure to people.

The owner had tried spending a lot of time lying on the floor and waiting for the rabbits to come over: they occasionally sniffed or butted her. When she approached them, the rabbits would move away. She picked them up once a month to clip their claws, and they needed several general anaesthetics every year to manage chronic dental disease.

The owner had acquired Emmett and Merrill because she had enjoyed spending time caring for her friends' rescue rabbits while they were on holiday. Her friend's rabbit would sit on her lap for long periods while she stroked it. She was frustrated that her rabbits would not engage with her in this way.

The owner was advised to use constructional approach training to decrease the rabbits' fear of her. She was told that her rabbits would probably never sit on her lap, but that she should be able to get to a point that she could spend time stroking and interacting with them. She understood that the bond between her and the rabbits would be disrupted each time that she picked up the rabbits, so found other ways to manage claw length.

It took several months for the rabbits to gain confidence with the owner. While the stress of the veterinary visits hindered the rabbits' trust in the owner, their behaviour did improve somewhat.

Many rehoming centres do not rehome rabbits with chronic dental disease because of the discomfort, pain and stress associated with this. This owner was not aware of the problems that she was likely to face when she rehomed these rabbits.

However, many owners interact with their rabbits in ways that are pleasurable for the owner but that cause distress or discomfort to their pets. To help owners move towards a more mutually rewarding relationship, they need to understand what their rabbits want. This section explores what 'good' and 'bad' interactions are from the perspective of the rabbit.

Allowing choice

With all interactions, animals should be able to express choice – if the interaction is forced, then the rabbit will be frustrated and is unlikely to enjoy the interaction. As well as being important for their welfare, allowing rabbits choice builds confidence and trust. However, this concept can be somewhat alien to owners who may be unaccustomed to reading signs of acquiescence and rebuttal. They may feel that they make important decisions for the pet's wellbeing that the pet does not appreciate ('cruel to be kind') or they may simply not see how they can give their pet choices.

At a simple level, owners can start to provide the ability to express choices by offering interactions and observing how the rabbit responds. If the rabbit wants to engage, it is likely to move forwards. A rabbit that does not want to engage may move away, may remain immobile or may shake the head or flick the paws.

If the rabbit does not interact, then the owner should move away and not offer the interaction again for a short period. This can be difficult for owners who have never done this before: their rabbit is likely to be wary, fearful and disinclined to engage. If an owner always picks their rabbits up to interact them, then when he stops picking the rabbits up (so they learn to trust him), he'll find that there is a period when he doesn't interact with the rabbits at all (he's not picking them up, and they don't want to come over). This can be disheartening. While transitioning on to a mutually pleasurable relationship, the owner is likely to go through a period that is considerably less pleasurable for them than their previous interactions. The rabbit will gain confidence and trust gradually, and then will start to interact more definitively and positively.

One case where this process is especially important is where the rabbit has learned aggressive behaviours to cope with fearful situations. Owners need to recognise that this behaviour has only been learned after two previous strategies (trying to escape, and remaining immobile) have been unsuccessful (see Diagram 3.7). Many older training texts will recommend that owners do not stop their behaviour if a rabbit shows aggression, as the rabbit learns that aggressive behaviour is effective. This logic is flawed: rabbits are aggressive because they have found that it is the only way to stop the owner's unwanted behaviour. If the rabbit learns that if it withdraws or stays still, it can avoid the unwanted interactions, it will no longer need to show aggressive behaviour.

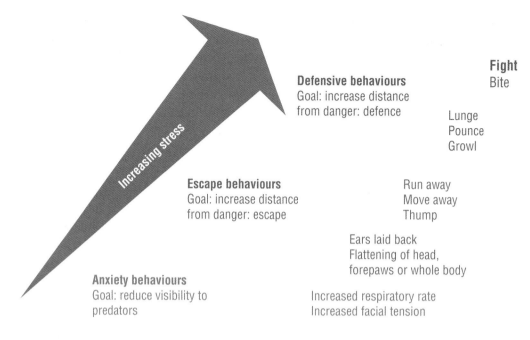

Fight
Bite

Defensive behaviours
Goal: increase distance
from danger: defence

Lunge
Pounce
Growl

Increasing stress

Escape behaviours
Goal: increase distance
from danger: escape

Run away
Move away
Thump

Ears laid back
Flattening of head,
forepaws or whole body

Anxiety behaviours
Goal: reduce visibility to
predators

Increased respiratory rate
Increased facial tension

Normal behaviours
Goal: satisfy non-fear motivations

Calm, relaxed
Interactive, exploratory

Diagram 3.7: Stress escalation ladder for rabbit behaviour.

Once owners have learned the signs of fear and distress, they need to act to reduce the stress of a situation before a rabbit resorts to aggressive behaviour. If the owner continues to act in a way that causes distress, while limiting the rabbit's options, then the rabbit may eventually stop showing the aggressive behaviour, but only because of learned helplessness. 'Giving up' makes the animal easier to handle, but will happen at the cost of the human–animal bond and the rabbit's welfare.

It is far more humane to prevent aggressive behaviour by reducing the motivation for the rabbit to show aggressive behaviour, rather than by trying to frighten the rabbit into submission. Advice on managing human-directed aggressive behaviour in rabbits is found in Section on Aggression at p. 163, but giving the animal choice is paramount to the successful resolution of the unwanted behaviour.

So, owners can give their rabbit choice over whether it wants to engage in potentially pleasurable interactions, such as grooming. However, there are other interactions between rabbits and owners that have very little chance of being pleasurable. These include being moved between a hutch and a run, or being put back into an enclosure after a period of access to a larger space. There are two options to increase choice in these situations:

◆ Owners can create an environment where the rabbit has the choice of when it wants to spend time in its hutch or run by connecting the two.
◆ Owners can train the rabbit to move between different spaces (or back into its hutch) by putting the behaviour on command.

Training, in essence, gives the animal a choice in which one outcome is rewarded. When the owner gives a cue, the animal can decide whether or not to respond to that cue. If it responds, it will be rewarded; if it does not, there is no active punishment, but equally no reward. Section on Training techniques at p. 141 will discuss how to train rabbits, but at this point, it is useful to point out that training is very good for an animal's welfare as it gives it some choice and control over its environment.

Mutually rewarding rabbit–human interactions

The following list details how owners can interact with their rabbits in ways that are satisfying for both parties. Bear in mind that these will only be pleasurable if the rabbit has some degree of choice over its interaction: if a rabbit does not want to engage with you, then even stroking it on the head may cause frustration or fear, rather than pleasure.

Mutually rewarding rabbit–human interactions include:

- Social grooming
- Resting in proximity
- Giving food
- Giving new toys
- Training

Social grooming

Owners who keep rabbits together will see social allogrooming behaviours. As previously described, allogrooming can be spontaneous or actively solicited (by lowering the head towards the other rabbit). Owners can mimic these behaviours by using similar signals, which gives the rabbit choice over how it responds.

Rather than going straight in to stroke the rabbit (especially if the rabbit is not actively soliciting attention), owners can 'ask' the rabbit whether it wants physical interaction by offering a closed fist near the ground, near the ground in front of the rabbit's face (Figure 3.47). If the rabbit wishes to be groomed, it will

lower the head. If the rabbit withdraws, or does not lower the head, it does not want to be stroked: this should be respected. If a rabbit frequently shows aggressive behaviour towards the owner, the owner should act to reduce the incidence of this behaviour before attempting this signal, as he or she is likely to be bitten on the knuckles.

One study found that male rabbits were three times more likely to display contact-seeking behaviour than females (d'Ovidio et al., 2016). However, this does not necessarily mean that female rabbits do not want social interactions: most will happily choose to interact despite showing fewer contact-seeking behaviours.

Owners should focus stroking, scratching and petting on the head and ears, and avoid touching the rabbit in less rabbit-acceptable places (Figure 3.48). Owners who have been bitten by their pets are likely to be cautious about touching the rabbits on the face, and so may be more inclined to stroke them on the back as it is further from the teeth! However, between rabbits, contact on the back tends to be sexually motivated, so is less acceptable to most rabbits if they expect to be groomed (Diagram 3.8). Owners will need to learn to read rabbit behaviour more carefully, but if they offer choice and respect the rabbit's choice, they can feel much more confident of stroking the rabbit on the face, which will be more rewarding for both.

Resting in proximity

Owners with house rabbits will often find that their rabbits choose to spend time in the room in which the owner is in, even if the owner is not interacting with the rabbit. As with rabbit–rabbit interactions, rabbits gain an advantage from spending time in large groups, as there are more animals alert to signs of danger. Therefore, merely being in visual contact with their owners can be rewarding to rabbits.

Figure 3.47: Owners can 'ask' the rabbit if it wants physical interaction by putting their fist in front of the rabbit's face (to mimic normal solicitation behaviour). A rabbit that wants to interact will lower its head.

Figure 3.48: Rabbits prefer to be stroked on the face and ears.

Diagram 3.8: Acceptable and unacceptable zones of contact on the rabbit. Green zones are preferred, red zones are disliked.

Giving food

Giving an animal food without asking for anything in return (giving 'treats') is very useful during the process of gaining or regaining their trust. They can learn that rather than being a source of fear, human hands can be a source of reward, and this will make training much easier for the owner.

As the rabbit becomes more confident with the owner, the owner can move on to using grooming as an interaction and food rewards for specific behaviours, such as recall. This will increase the response of the rabbit when behaviours are required, and helps the owner to enjoy a relationship with their rabbit that is not entirely dependent on food. If an owner wants to give a rabbit a piece of vegetable or some leaves, even just calling them over (practising a recall comman) will give the rabbit more choice and reward desirable behaviours.

Giving new toys

Rabbits do not play with toys in the same way that humans or dogs do. As previously described, there are two types of play behaviours: specific and functional. Rabbit specific play behaviours are primarily social (chasing, displacing other rabbits) and are barely present in adulthood. Rabbit functional play behaviours are mostly locomotor, such as short sprints or binkying.

Rabbits kept with other rabbits are unlikely to interact with toys for a long time unless there is a good motivation to do so. When a novel object is placed in the environment, most rabbits will investigate it to ascertain whether it poses risk or offers reward. If the former, they will avoid it, if the latter, they will engage more with it, and if neither, they are likely to ignore it unless it is impeding access to another resource, when they will attempt to move it out of the way. Rabbits kept in social isolation may be more likely to engage with less interesting toys because they are bored and lack social stimulation.

Toys that will be reinforcing to both the rabbits and to the owner (who will feel rewarded for their purchase if the rabbits engage with it for a period of time) need to comprise a food reward: either because the whole item is consumable (tree branches) or because food can be hidden within it (such as a puzzle feeding ball). Rabbits detect food through scent, so the food does not need to be visible. Destructible, manoeuvrable toys, such as scrunched cardboard or paper containing food rewards (Figure 3.49), allow rabbits to show a range of normal behaviours (biting and digging) in the process of accessing the food.

Training

Training is a very good way of strengthening the human–animal bond as it is reinforcing for both the trainer and the animal. For the rabbit, the process of learning is stimulating and reinforcing, and the owner becomes a source of food rewards and fun. For the owner, watching a rabbit respond to training is reinforcing, and increases positive feelings towards the rabbit. Additionally, putting behaviours on cue may help owners to avoid situations that they find more stressful (such as catching the rabbit).

Aversive rabbit–human interactions

Aversive rabbit–human interactions include:

- Owner picking up rabbit
- Owner restraining rabbit
- Owner 'trancing' rabbit
- Owner frustrating rabbit

Figure 3.49: Hay can be provided in a number of different forms to stimulate investigative and consumptive behaviours.

Owner picking up rabbit

Over the last few centuries, humans have come to view rabbits as pets that are frequently lifted and held, partly because their fear response of 'freezing' means that they may not try to escape, and partly because their size means that humans need to crouch on the floor to interact with them at their level. Children's books and media materials have helped to perpetuate this societal attitude, with most images of interactions between humans and rabbits depicting rabbits being held.

Animals that themselves have dextrous 'hands' (such as monkeys, apes and humans) are likely to have instinctive positive feelings about being picked up. This is because, when they are infants, members of their own species frequently pick them up. Humans interact with their world primarily with their hands: mothers lift their children to show affection or protection; friends greet each other with a hug. Rabbits don't have hands, so are therefore not picked up in this way by other rabbits – being lifted off the ground is extremely stressful. This creates a discord between how humans and rabbits want to interact (Bradbury and Dickens, 2016).

It can be hard for owners to recognise that lifting their rabbits causes stress. The signs of stress can be very subtle, and may not be noticed by the owners: rabbits, unlike dogs, cannot kill or badly injure humans when stressed or scared. To many owners, this makes the problem of inappropriate handling seem less pressing. Some rabbits may show fear through growling, lunging or biting; others may freeze or learn to tolerate being picked up. Owners may interpret the 'freezing' behaviour when lifted as acquiescence or willingness, rather than fear or learned helplessness. Subtle signs of fear or discomfort (facial tension, pupil dilation, abdominal flattening response) are often overlooked. Owners may notice that their pets are less likely to interact with them – a pet rabbit that is frequently picked up is less likely to voluntarily approach the owner (Mullan and Main, 2007).

When the human–animal bond is strong, owners are more motivated to meet the various welfare needs of their pet. If owners pick up their rabbits frequently, the human–animal bond inevitably suffers. Teaching owners how to correctly interact with their rabbits will strengthen the bond, providing more motivation for the owner to improve rabbit husbandry in other areas. The perfect human–animal bond is mutually rewarding: it should not be imposition by one and tolerance by the other.

In some cases, where the husbandry is poor, rabbits are more likely to tolerate being lifted. In laboratory rabbits, it has been shown that rabbits kept alone are less likely to struggle when lifted, perhaps due to the social deprivation in a socially dependent species (any form of contact is preferable to social isolation). This can make it difficult when advising owners to seek a companion for their single rabbit if their major form of interaction with the rabbit is by picking it up. However, the welfare gain from having a companion is greater than the welfare gain from stopping lifting the rabbit, so always address the larger husbandry problem first.

There are two ways to reduce the stress caused by picking up rabbits. The first method is mostly under the control of the owner; the second method involves breeders, pet shops, owners and veterinary staff.

These methods are:

- decrease the number of times that the rabbit is picked up
- ensure that, when it is absolutely necessary to pick up the rabbit, the stress is minimised.

Decrease the number of times that the rabbit is picked up

There are two major reasons for owners to pick up their rabbits: practical reasons (putting the rabbit into the hutch, checking the rabbit's health) or emotional reasons (to show affection for the rabbit).

If owners are to stop picking up rabbits for practical reasons, they will have to change the rabbit's environment so they no longer need to catch the rabbit. Runs attached to hutches (as advised by the RWAF) eliminate the need for rabbits to be lifted up. Rabbits can be easily trained to recall to a whistle, to go into their cage or hutch on command and to do simple tricks for food rewards. Together, these allow the rabbit some freedom of movement while reducing stress, not only for the rabbit, but also for the owner, who no longer has to catch the rabbit to return it to its enclosure. Many owners would prefer not to have to catch their rabbit frequently, but the major barriers here are likely to be around the environment.

Owners also lift rabbits to perform important regular health checks, as recommended by vets. However, there are much less invasive ways of checking health that do not require lifting (see earlier section on Health at p. 52). Owners can look for changes in normal behaviourrs, like feeding, activity, and resting postures. They can assess health by training behaviours that allow them to look for clinical signs. They can look at the claw length and palpate the head for injuries, lumps, or painful areas. These things can be assessed as part of a pleasurable interaction with the rabbit, without needing to pick them up. Owners can visually check the anogenital area after using a food reward above the animal's head to encourage it to stand up on its hind legs. Needless to say, this should be done on the floor. In this situation, it is education that is the key to reducing the frequency of the rabbit being picked up.

If an owner picks up the rabbit to demonstrate or to receive affection, then it is much harder for them to stop this behaviour. They will have to learn mutually satisfying ways of interacting with their rabbit. If they are not used to reading their rabbit's behaviour, then this can seem a significant challenge.

Additionally, it may be hard for them to change their behaviour – if they accept that their rabbit does not like being picked up, then they may feel some degree of retrospective guilt for their previous actions. These owners need empathetic education about rabbit behaviour and rabbit motivations – acknowledging that they were following bad advice correctly shifts blame. They need advice on suitable interactions (head lowering as solicitation of interaction, suitable areas for stroking, reward-based training and others).

Don't underestimate the difficulty of changing the owner's behaviour here. It might be useful to suggest that the owner attempts a 'trial period' where they don't pick up the rabbit for two to four weeks, rather than an abrupt and complete cessation, which the owner may find very difficult emotionally.

Ensure that, when it is absolutely necessary to pick up the rabbit, the stress is minimised.

There are two strategies for minimising stress during handling: better socialisation of rabbit kits, and more humane handling of adults.

1. Better socialisation of rabbit kits
   Rabbit kits below the age of seven or eight weeks have a much reduced fear response, and can learn to tolerate human handling. At this age, the rabbit is with the breeder rather than the new owner. Rabbit breeders can help to reduce later stress in two ways. First, by selectively breeding from rabbits that are calm and confident around

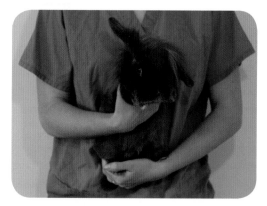

Figure 3.50a–b: When rabbits need to be picked up by human hands, they should be held firmly with their paws supported.

humans. Second, by ensuring adequate socialisation of young rabbits: either by gently picking up the kits, or even by exposing the neonatal (newborn) rabbits to the scent of humans (Dúcs et al., 2009). Both of these methods will mean that in adulthood, if there is no option but to pick up the rabbit, it will experience less fear and will suffer less.

2. More humane handling of adult rabbits

If a rabbit must be lifted, the handler should support the rabbit so it feels secure. This includes supporting the paws to allow the rabbit to brace and stabilise itself against the movement (Figure 3.50). Rabbits should never be lifted by the ears, or by the scruff alone: neither part of the anatomy is strong enough to take a rabbit's weight without causing pain and injury. Supporting the rabbit's thorax and forelimbs with one hand while holding the hind paws in the other, or supporting the rabbit's back end, with all of its paws on your chest, seem to be the least-worst solutions. When lifting is required, containing the rabbit in a towel prevents direct contact of human hands with the paws, which seems to be especially distressing (Figure 3.51).

To give the rabbits more control over their situation, owners can train rabbits to voluntarily enter a box or sit on a towel to be moved. Advice Sheet 13 gives a guide to training this behaviour. It requires a lot of trust from the rabbit in the owner, and so represents an advanced level of training. This should only be recommended to owners who already provide good husbandry for their rabbits and have trained several behaviours: if not, the owner is very unlikely to succeed.

However, regardless of correct lifting styles and socialisation, owners must recognise that to improve a rabbit's welfare, the rabbit should be picked up as infrequently as possible.

Owner restraining rabbit

Even if rabbits are not lifted, they dislike being restrained or confined, a motivation that is important for survival in the wild. Restraint includes being confined in human hands (even if not being lifted), but also being trapped in a small space. Rabbits prefer shelters with multiple entrances and exits, so they have routes of escape if needed (Figure 3.52). When a rabbit is restrained, the routes of escape are cut off, depriving the animal of its major coping strategy: running away.

Vets may advocate using the 'bunny burrito' method of restraint for simple procedures, such as syringe feeding or claw clipping. This involves wrapping the rabbit tightly in a towel. This is an

Figure 3.51: It is usually easier to train rabbits to be lifted in a towel rather than lifted by human hands, suggesting that this experience is less unpleasant.

Figure 3.52: Rabbits prefer shelters with multiple entrances and exits. Owners can make these from boxes or repurposed children's toys.

appropriate form of restraint for simple procedures (it avoids contact with human hands, but still provides sufficient immobilisation), although it is nevertheless unpleasant and should only be used when necessary.

As discussed in the previous section on mutually rewarding interactions, owners should always give their rabbits a choice over whether or not the animal wants to interact with them. If the owners are behaving in this way, the rabbit should not need to be restrained.

This motivation explains why owners can find it very difficult to train a rabbit to go into a transport cage on demand: in addition to any negative associations the rabbit has with the box (learned through previous transport experiences), the very nature of the small, confined space is unpleasant to the rabbit. Rabbits can be trained to do this, but the food rewards need to be highly valued, and the training, once learned, will need to be reinforced at regular intervals

if the behaviour is to be retained. It is easier to train this behaviour if the transport cage forms part of the rabbit's familiar environment (Figure 3.53).

Owner 'trancing' rabbit

Tonic immobility, 'hypnotising' or 'trancing', refers to the state of paralysis that is induced in a rabbit when it is turned on to its back. It is thought to be a way of deterring predators – the rabbit 'plays dead'. Some rabbits remain immobile on their back even when no longer held. Some do not become immobile at all. As the rabbit seems 'relaxed', some owners think that the sensation is pleasurable: hence the popularity of internet memes of rabbits on their backs. This is not true: research has shown that the rabbits are aware of their surroundings, are exhibiting a fear response and can still feel pain.

One study (McBride et al., 2006) found that rabbits found the whole procedure of

Figure 3.53: If rabbits are to be trained to go into a box on command, the box should be part of their environment for a while to desensitise them to its presence.

tonic immobility very aversive. When the rabbits were turned on to their backs, they showed typical fearful behaviours (flattened ears, widened eyes, increased muscle tension and overt struggling). While the rabbits were tonically immobile, the breathing rates and heart rates increased, as did the plasma levels of corticosterone, a stress hormone. After the procedure, the rabbits were more likely to groom themselves and hide, and less likely to explore, indicating that they were recovering from the procedure. This suggests that there is a strong aversive emotional response caused by being tranced.

Owners should never trance rabbits, and veterinary surgeons should only use this procedure if there is no other option available.

Owner frustrating rabbit

Rabbits will show frustration behaviours if they are 'teased', for example, if food is offered and then withdrawn, or if the owner repeatedly exposes the rabbit to something it dislikes. Typical frustration behaviours are head-shaking, growling, lunging, pouncing or biting at the disliked object, or running away. Owners may find these behaviours entertaining, or see them as part of play, but they increase the rabbit's stress and increase the chance of agonistic behaviour if the owner then tries to interact with the rabbit, as the frustration is redirected. It's useful to discuss how the owner plays with the rabbit, and discourage interactions that cause frustration as they may indirectly affect other behaviours.

Rabbits and children

'Rabbits are ground-loving prey animals, who become friendly and responsive when properly treated. But rabbits are vulnerable to injury if handled badly and rarely appreciate being cuddled. Therefore, rabbits do not make good children's pets, but can make successful family pets if parents respect the needs of the rabbit and the limitations of the children. Adults must accept all the responsibility of caring for the rabbit.'
Rabbit Welfare Association & Fund policy statement, 2014

Rabbits are often seen as children's pets. Many children's stories feature children holding and cuddling rabbits, and this image is still often used in marketing and advertising. It is hard for children to read the subtleties of rabbit behaviour – signs of stress, fear and disease are often not obvious. Therefore, acquiring a rabbit as a child's pet will often cause frustration for the child and suffering for the rabbit. The RWAF issued the above policy statement to reduce the number of rabbits given up for rehoming by families dissatisfied with their 'child's pet'.

Rabbits can make good family pets, providing certain rules are followed. If a pair of rabbits is to be introduced into a family, they should be the responsibility of the adults, not the children. The children should be taught to let the rabbits choose to interact with them, to never chase them and to never pick them up.

If the owners have a baby, there are various ways that they can increase the likelihood of the rabbits and children developing a good relationship. Babies disrupt normal routines and make loud noises, which can easily scare rabbits. The rabbits must therefore be able to move away from the baby when they are scared. When the baby is calm and relaxed, the owners should put rabbit food around the baby so the rabbits form pleasant associations with him or her (Figure 3.54). The owners should also reward the rabbits

Figure 3.54: Scattering food around a baby allows the rabbits to form pleasant associations with it.

children's attention on to another activity reduces their frustration.

> *Explaining the rabbits' actions helps children to empathise with the animals and to learn what they like.*
>
> 'When she grinds her teeth like that, she's feeling happy and relaxed. She really likes it when you stroke her gently on the face and ears.'
>
> 'Can you see how the rabbits are snuggled together and sleeping? When they look like that, they don't want to be disturbed.'
>
> 'When he moves away from you, he doesn't want to play with you any more. He was scared because you moved very suddenly. Next time, try to move more gently and he is less likely to run away.'

for showing exploratory behaviours around the baby, rather than fear behaviours. As the child grows, they can be taught to gently stroke the rabbits on the head and ears – teaching them to use the back of their hand reduces the temptation to poke or grab.

Young children often find it hard to interact with pet rabbits. Their movements are often sudden and so the rabbits are likely to be scared. If the rabbits run away, it is hard for the children to resist the temptation to chase them, and they can be rough and may injure the rabbits during handling. Parents should only allow contact between the children and rabbits when the children are tired and calm. Young children can be taught to only stroke the rabbits when they are lying on their front at the same level – this helps them to see the world from the rabbits' point of view, and, more practically, prevents them being able to give immediate chase! The parent should stop the interaction before the rabbits are bored to prevent the child feeling frustrated when the rabbits move away. Redirecting the

As the children become older, parents should explain what the rabbits' behaviours indicate. This helps them to see the rabbits as individuals with emotions, and helps them to learn how to modify their own behaviour to encourage the behaviours they want in the rabbits. Parents should teach the children to recognise signs of fear, and to respect the rabbits' choice to interact or to leave them alone. Parents should also teach the children how to encourage the rabbits to interact more with them, so they learn 'good rabbit manners' rather than just learning to avoid 'bad rabbit manners'. When friends visit, children can be tempted to try to force the rabbits to interact with their friends – but giving them responsibility to teach their friends how to interact with the rabbits in acceptable ways can help to reinforce good behaviour patterns.

Even quite young children can hand feed rabbits, and this provides a rewarding interaction that does not involve contact. They can scatter pieces of vegetable on the floor to watch the rabbits search for them, and they can point to pieces that the rabbit has missed (unlike dogs, rabbits do not instinctively follow the direction of a point, but this behaviour can be trained). If the rabbits have good recall, children can run to different parts of the garden and call them over to give them pieces of concentrate food. Older children can be taught how to train the rabbits to do different behaviours – often showing more application than their parents!

The only physical contact that children should have with the rabbits is stroking them on the head and ears. Children should be taught to offer a closed fist in front of the rabbit's nose – if the rabbit lowers its head, they can stroke it, but if it doesn't, they should move away. This sort of impulse control is hard for young children, so they should always be supervised and distracted away if the rabbits do not want to interact.

Children and rabbits do not automatically form mutually beneficial relationships. If adult owners invest time and effort in the learning process, then children can learn to have close relationships with rabbits, and the impulse control developed will serve them well in different parts of their lives.

Interactions with other pet species

Many rabbits are kept in households where there are pets of different species. Depending on the husbandry of the rabbit, the extent to which the different species interact may vary. House rabbits may live in close contact with predator species, and, if introduced appropriately, these interactions may be rewarding and stimulating for the rabbits. However, inappropriate management of multi-pet households can cause significant stress and risk to rabbits.

Many of the considerations previously discussed for rabbit–human interactions are equally applicable to other rabbit–non-rabbit interactions. Rabbits have a requirement for same-species companionship: companionship with an animal of another species can never be as rewarding as companionship with another rabbit. This is because the different species have differing social requirements, communication strategies and preferred interactions. However, providing that the rabbit is safe (i.e. no predatory behaviour is shown by the other species), the animals can learn to enjoy each other's company in ways that are mutually rewarding.

Prior to routine neutering of rabbits, rabbits were often kept with guinea pigs: allowing social contact for both species without risk of reproduction. However, the welfare of both animals can suffer as their needs differ. Guinea pigs have a requirement for external sources of vitamin C, so should not be fed on the same diet as a rabbit needs. The communication behaviours are also different – rabbit–rabbit social pairs spend much more time grooming each other than do pairs of guinea pigs, and the grooming behaviour is solicited in different ways. Failure to understand these communication behaviours can lead to frustration and stress on both sides.

As rabbits are bigger than guinea pigs, frustration-related agonistic behaviour from a rabbit can do significant damage to a guinea pig. The rabbit may also attempt to guard resources from the guinea pig: the lack of common social behaviours prevents the normal resolution of this conflict. Finally, unneutered rabbits or guinea pigs may show sexual behaviour towards their companion, which can cause frustration, distress or even injury. Where rabbits and guinea pigs are currently kept together and where the relationship is amicable, they should not be separated, but owners should never be advised to introduce animals from these two species with the aim of keeping them together.

Mutually rewarding interactions between rabbits and other species

Rabbits will find interactions with other species most rewarding when they are similar to inter-actions with other rabbits. Therefore, desirable interactions are similar when a rabbit interacts with any non-rabbit species – humans, dogs and cats:

- allogrooming
- resting in proximity
- investigating in proximity.

Allogrooming

As previously discussed, this is extremely important for a rabbit's wellbeing. Some non-rabbit species groom each other more than others: cats are more likely to groom a rabbit than are dogs, although some bitches will lick a rabbit as they would a puppy.

Rabbits may learn when they are likely to be groomed (by a cat that is washing itself, for example), and the other animal may learn to respond to rabbit behaviours soliciting grooming. If this occurs, it is very beneficial for the rabbit.

Resting in proximity

As previously discussed, visual contact with other animals (providing they are not a source of fear) is rewarding, so rabbits will often follow other pets around the house space and sleep nearby.

Investigation in proximity

Other animals can be a source of stimula-tion and enrichment to rabbits, and being in visual proximity when exploring a new area

or a garden will allow the rabbit to show more investigatory behaviour.

Aversive interactions between rabbits and other species

Aversive interactions are:

- frustrated sexual behaviours
- experiencing predatory behaviours
- scent contact with predator species.

Frustrated sexual behaviours

If unneutered animals are kept together and have no normal outlet for sexual behaviours, these behaviours can be redirected on to an animal of another species. This can cause pain, distress and learned helplessness: the motiva-tion of an animal to reproduce is very strong and it may be hard for the unwilling focus of the sexual attention to escape. The pursued animal may try to defend itself and cause injury to the animal trying to mate: this can be espe-cially damaging when the latter mounts the front, rather than the back. Injuries including abdominal laceration and penile amputation have occurred.

If animals are kept together and unneutered, owners must be very vigilant to the signs of sexual behaviour directed at another pet. If this occurs, the animal should be neutered and sep-arated until the hormone levels have subsided (typically one to two months).

Experiencing predatory behaviours

Predatory behaviours, such as stalking, pouncing or biting, indicate that the preda-tor species views the rabbit as potential prey. This is extremely serious. The animals should be completely separated. It is extremely dif-ficult to change the motivation of a predator species to hunt: this is one of its strongest motivations. The rabbit will be at risk of injury or death.

Scent contact with a predator

Some rabbits instinctively dislike the scent of certain species. The smell of ferrets, for example, can provoke either escape behaviours or strong aggressive responses – an understandable adaptation to avoid predation. Desensitisation and counter-conditioning, therefore, may be ineffective, and long-term exposure to these scents may cause significant stress. Owners who frequently handle the predator species before handling the rabbits may find that their rabbits are unduly fearful or aggressive towards them. Therefore, when you're doing a behaviour consultation, always find out if the owner keeps other pet species.

Introducing another animal

When introducing another animal into a household with a rabbit, the owners need to ensure that neither the rabbit nor the other animal is injured or distressed. While an owner's first priority should be the pets that they currently own, many owners will not think about this responsibility. Introductions of new animals can cause significant distress to the pets already within the house. Rabbits are very susceptible to stress, which can lead to gastrointestinal ileus, and may cause death.

When introducing a new pet to a home that contains rabbits, the owner should consider the following questions.

- Does the evolutionary background of the new animal include predatory behaviour towards small mammals?
- Will the new animal be expected to share the same space as the rabbit, or will it be kept completely separately?
- To what extent does the owner need to 'trust' that the new pet will not harm the rabbit?
- What is the age of the new animal?

- Has the animal shown previous predatory behaviour?

Let's explore these in more detail.

Does the evolutionary background of the new animal include predatory behaviour towards small mammals?

Any predator species that has evolved to hunt and kill small mammals will retain the same predatory motivations. It is possible to overcome these motivations in a variety of different ways: exposure to the rabbits when the predatory animal is young (so rabbits are seen as conspecifics rather than prey); positive reinforcement for ignoring the rabbits; or selection of animals with a size disparity that is likely to discourage engagement (a large rabbit is less likely to be perceived as 'prey' by a small cat. However, do not apply this logic with ferrets!).

Additionally, within a species (especially within dogs), different breeds have been selected for very different motivations. It is much easier to train a collie (which has been bred to herd sheep) to ignore rabbits than it is with a terrier (which has been bred to kill small mammals). Cats that are from feral lines (where hunting has been advantageous in recent generations) are likely to have a higher prey drive than breeds that have been selected for many generations for a certain aspect of their appearance.

Owners should be aware that many adult dogs and cats might never be able to be trusted not to display predatory behaviour towards a rabbit. Additionally, it is not possible to always predict the likelihood that an animal will show predatory behaviour until it is in proximity with a prey species (many dog owners will believe that their dog would never attack another animal until they are faced with painful evidence to the contrary). If the owner has taken on an animal that has a high prey drive, then the owner has an obligation to ensure that the new animal can never gain access to the rabbits. They should also prevent the animals being in eye contact, as this will be frustrating to the dog or the cat and will continually induce fear in the rabbit.

Will the new animal be expected to share the same space as the rabbit, or will it be kept completely separately?

Owners need to set appropriate expectations when housing different species in close proximity. If the owners have house rabbits, then any dog or cat introduced will need to be less reactive to the rabbits than if the rabbits were housed outside in a hutch and run complex.

To what extent does the owner need to 'trust' that the new pet will not harm the rabbit?

The general guidance is that rabbits should never be left alone with predator species: changing motivations of the predator may result in an unforeseen attack. However, some rabbits are successfully kept with cats and dogs (Figure 3.55), especially if either the rabbit or the other pet has been introduced as a juvenile when its fear responses and social behaviours were still being learned.

House rabbits, despite sharing the same space as other pets, need not be in close proximity all of the time. Owners may separate the animals in different rooms when the owners are out of the house, so the animals are never unsupervised. If owners leave the animals together when they leave the house, there should be many spaces that the rabbits can access but the larger pets cannot – this provides protection if needed.

If an owner seeks advice before introducing a new pet into the household, you should encourage them to state their expectations around the degree of contact that they are expecting between their animals, and encourage them

to consider ways to protect the welfare of all of their animals.

What is the age of the new animal?

There are advantages and disadvantages to introducing animals of different ages. Young rabbits have decreased fear responses and increased exploratory behaviours, so are less likely to show the escape behaviours that trigger chase behaviours in predators. However, they are relatively smaller, making them an easier target.

If an adult rabbit is introduced to a puppy or kitten, the rabbit will seem relatively larger to the other animal, which may decrease the likelihood of it being viewed as a prey species. However, puppies and kittens show a lot of play behaviours, some of which is practice predatory behaviour, and may cause fear to the rabbits. Young predators have also not learned impulse control, so it may be harder to encourage them to leave the rabbits alone. Their continuous attempts at interaction can cause a lot of distress.

Additionally, be aware that the motivations of the predator species may change with age: even if the puppy behaves calmly around the rabbits, there may come a point (triggered by an external stimulus, a behaviour of the rabbit or an internal stimulus in the other animal) where it shows predatory behaviour towards the rabbit. Young animals should always be supervised.

Young rabbits can be introduced to a suckling bitch or queen as part of her litter. The aim of this is that the rabbits are seen as her offspring, and therefore will not be a focus of predatory behaviour from either the mother or the litter. At this point, the hormonal influences reduce predatory drive, and the rabbits, if the scents are mixed, can be well accepted. However, the rabbits' behaviour will differ substantially from the young kittens or puppies:

Figure 3.55: If appropriately introduced, rabbits and dogs can form stable relationships. Photo: Noémie Maigret

this can be a source of stress for the dam, and she may repeatedly move them into the litter of puppies, prevent them leaving and interact with them in ways that they don't understand. This can be a problem if the rabbits are prevented from showing normal behaviours (such as caecotrophy) and accessing a normal diet.

Has the animal shown previous predatory behaviour?

This is not relevant for a young puppy or kitten, but gives information on the likely response of the adult animal to a rabbit. Dogs that like to chase birds or squirrels, and cats that bring home birds or small mammals, are very unlikely to be good candidates for introduction to rabbits, even after significant training.

Specific methods for introducing a dog

There is a risk to any rabbits that are introduced to a dog. Owners need to be aware of this risk and take steps to minimise it: by selecting a dog that is unlikely to have a prey drive and by ensuring the dog is suitably restrained. The goal is to build confidence in the rabbits, so they do not run away – if they do, this behaviour is likely to trigger the dog to chase the rabbits.

Introducing a puppy to rabbits

The goal, when introducing a puppy to rabbits, is that the puppy never becomes excited. A high arousal state reduces the puppy's ability to control its impulses, making it more likely to show behaviours that will make the rabbits run If the rabbits do run away, they will be stressed and the puppy will be excited. If the puppy is once allowed to chase the rabbits, the reward of this behaviour may be so great that the puppy can never be trusted around rabbits. This needs to be avoided at all costs.

Therefore, if a puppy is to be introduced to rabbits, it needs to have a basic level of training beforehand. It must sit and lie down on command, and it should also have been trained to 'leave' a toy or food (when the owner gives the command, the puppy should look away or move away from the toy or food, whereupon the owner reinforces the behaviour by giving a higher value reward). It should also have been accustomed to wearing a collar and a lead.

It is good practice to ensure that the puppy is tired before meeting the rabbits: an enthusiastic game for half an hour with a lot of running around should ensure that it is less reactive to stimuli. The rabbits should be in a cage or behind a barrier (with places to avoid direct visual contact from the puppy), and the puppy should be on a lead to prevent it going straight up to the cage.

The owner should enter the room or approach the outside cage with the puppy, and almost immediately ask the puppy to perform a cued behaviour ('sit' or 'down'). The behaviour should be reinforced. After several repetitions of the trained behaviour, the owner and puppy should take a step closer to the rabbits and repeat this sequence. Initially, the puppy may well not notice the rabbits (perhaps for several training sessions), especially if they are not moving: this is a good outcome, as the puppy is desensitising to the smell of the rabbits. If the puppy shows any signs of excitement (tensing, staring, whining, barking or lunging), the owner should lead the puppy out of the room. If not, the session should last about five minutes before the puppy is taken out of the room and the training is stopped.

These short training sessions should be repeated at any point that the puppy is tired.

Food should be scattered in the cage for the rabbits during these training sessions. The goal is to approach to a distance where the puppy can still focus on the training, rather than on the rabbits. Once the owner has managed to approach the cage so the puppy is next to the cage and still able to perform cued behaviours, the owner should ask the puppy to lie down (and continue giving rewards for this behaviour) while rewarding the rabbits for any movement towards the puppy.

When the puppy can lie next to the cage without showing undue interest, it can be allowed to sit and look at the rabbits. Use the 'leave' command to get the puppy to look away from the rabbits to get a treat. The owner can increase the intensity of the stimulus by recalling the rabbits from one end of the cage to the other (encouraging them to move) and asking the puppy to 'leave' the rabbits.

If the puppy can stay calm at this point, the cage can be opened or barrier removed and the rabbits allowed to come forwards to investigate the dog (on their own terms: this may take several training sessions). The puppy should be reinforced for remaining in a 'down' position, or encouraged to recall to the owner away from the rabbits.

If at any point during these sessions, the puppy shows signs of excitement or the rabbits show signs of fear, the puppy should be removed from the room immediately. Do not punish the dog if it shows these behaviours: this will increase fear and arousal and make it much harder to train relaxed, calm behaviours. As discussed above, young dogs (under a year of age) should never be left alone with rabbits.

Introducing an adult dog to rabbits
When introducing an adult dog to rabbits, it is essential that the dog cannot harm the rabbits

(the risk of immediate, permanent harm to the rabbits is greater than with a puppy). The dog should not be punished for negative interactions with the rabbits, but these must not be rewarded (i.e. the dog should be on a lead so if it lunges at the rabbits, it cannot catch them – the behaviour is not rewarded). The dog should instead be positively reinforced for ignoring the rabbits. Although the principles of training are similar to those described in the previous section, it can be much quicker to train a well-behaved adult dog to ignore rabbits, as it has substantially better impulse control.

When choosing a dog to introduce, bear in mind the caveats described before (breed, history of predatory behaviour, degree of training). If the rabbits are caged, the owner should use the training sequence described in the previous section. However, a well-trained adult dog can be introduced to free-range house rabbits, as described below.

The set-up of the environment in which the animals will meet is very important. The dog should be lying down (so it seems smaller and less threatening). Letting the rabbits choose how closely they approach the dog will reduce their fear. If they recall to a whistle, they can be called over towards the dog. When the rabbits approach, the owner should give food to both the dog and the rabbits, scattered on the floor to promote investigatory foraging behaviour. Do not attempt this with very hungry or food-possessive dogs, as they may snap at the rabbits in competition for the food. Grass is a very useful foodstuff for this practice as it is of very high value to rabbits and very low value to dogs.

Once the animals have eaten the food on the floor in proximity to one another (non-associative learning, see next section), the owner can then start to reinforce desired behaviours from either the rabbits or the dog.

Desired behaviour from rabbits

Desirable behaviours include any investigative behaviour or any behaviours that indicate a relaxed state. Dogs are much less likely to react to rabbits that are approaching them or ignoring them than those that are running away. If the rabbits are clicker trained, they should be clicked and rewarded (see section on clicker training at p. 143) for any steps towards the dog, for interacting with each other, or for resting or washing themselves.

Desired behaviour from dog

Desirable behaviours include any form of retreat from the rabbits: this indicates a good level of impulse control. If the dog is clicker trained, the owner should click and treat the dog when it looks away from the rabbits or when it makes eye contact with the owner. The outcome of reinforcing these subtle behaviours is usually that the dog, after a period, gets to its feet and moves away from the rabbits and towards the owner: this behaviour should receive a substantial reward.

Owners should try to keep the dog's arousal level as low as possible: exuberant play behaviours or barking are likely to startle the rabbits, making them run away, which may trigger an unwanted response from the dog. Once the rabbits have shown fear towards the dog, and the dog is aware that they will run (which is enjoyable for the dog), it is possible, but very hard, to train the animals to accept each other without showing these behaviours.

If a dog stalks or lunges at the rabbits, the lead should prevent it from injuring them. Do not punish the dog, but remove it from the situation. These behaviours indicate that the dog should never be trusted alone with the rabbits.

Introducing a cat

It is not uncommon to see households where cats and house rabbits are kept together amicably. Cats can learn to groom rabbits when solicited, which is very rewarding for both animals (Figure 3.56). A poll on the House Rabbit Society Facebook page found that the majority of cat–rabbit households reported that the animals lived together calmly without agonistic interactions (see Diagram 3.9). Nevertheless, cats are predatory, and great care should be taken that they do not injure or kill the rabbits.

Figure 3.56: Rabbits and cats can form strong bonds, providing that the cat does not see the rabbit as potential prey. Photo: ~Liz. 'BFFs'. Flickr.

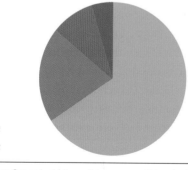

- ■ Cat and rabbit coexist peacefully, not necessarily best of friends
- ■ Rabbit is aggressive towards cat
- ■ Cat and rabbit have close relationship
- ■ Cat is aggressive towards rabbit (kept separately)

Diagram 3.9: Rabbit–cat relationships in the home (as reported on the House Rabbit Society Facebook page).

The goal of introducing a cat to rabbits is that the rabbits feel confident and the cat does not feel predatory. Pick a time where the cat is likely to feel calm and sleepy. Some recommendations suggest that the cat should also have recently been fed – this may help, but the predatory drive in cats is independent of hunger (Adamec, 1976).

It is usually easier to introduce a cat on territory in which the rabbit is established. As previously discussed, rabbits defend territory against incomers, and so are more likely to show confrontational approach behaviour rather than fear-based retreat. Cats can escape to higher surfaces, so are unlikely to be hurt by the rabbits, but inducing an anxious or fearful response in the cat initially reduces the likelihood that the rabbits will be seen as potential prey. If rabbits are to be introduced to a resident cat, then it is wise to introduce large breed rabbits: cats are more likely to feel predatory towards small prey.

The ideal combination of these animals is a calm cat and assertive rabbits. The owner should be ready to interrupt if the cat shows any predatory behaviour: either by catching the cat, or by squirting it with water if it cannot be caught. Cats can be introduced to caged rabbits, but the response may be skewed towards a predatory one: the cat is unlikely to be made fearful (as the rabbits cannot approach it) and the rabbits are more likely to be fearful (as they do not have choice over their interactions). A good first interaction is that the rabbit approaches the cat and the cat retreats: this reduces the prey motivation of the cat.

The situation is slightly different with kittens: the size difference means that most rabbits will be larger than the kitten, so the prey motivation will be reduced (Figure 3.57). Additionally, kittens may find it harder to escape, so could be injured by a territorial rabbit. It may be easier to

Figure 3.57: Kittens are smaller than most adult rabbits, so may be better accepted, but owners must monitor for predatory behaviours as the cats become adult. Photo: ~Liz. Flickr. 'Bunnies have bonded'.

introduce a rabbit to a kitten on neutral territory, with places for the kitten to hide, or introducing them in adjacent cages so they can get used to each other before being introduced in a wider environment. As discussed earlier, always supervise rabbits and young animals of different species as their motivations may change through adolescence and into adulthood.

Affective experience domains

The last sections have explored the rabbit requirements in terms of nutrition, environment, health and behaviour, and identified those features that lead to positive or negative affect. The rabbit's mental state can be described as the cumulative effect of positive or negative affects within each of the previous domains – this is the final 'domain' of the Five Domains framework.

When performing your behavioural assessment, it can be useful to frame your findings around this model. What is the diet of the rabbit like? Therefore what are the positive and negative affects (feelings or emotions) that it is likely to experience? When all of the domains are considered, what are the major negative affects that are likely to arise (loneliness,

restricted space to show normal behaviours)? How can the owner give the rabbit a better environment to increase the positive affects that it experiences?

The balance of the positive and negative affects may not be clear, but it is useful to identify negative affects, or lack of positive affects, that may be contributing to an animal's behaviour. This information forms part of your behaviour modification plan.

The balance of an animal's positive and negative experiences influences its mental state. In this section, we'll explore how a rabbit's emotions and personality affect how it perceives its situation.

Why do individual rabbits have different reactions to the same stimulus?

Positive and negative affects are the same between different animals – for example, all animals dislike the feeling of fear. However, that feeling may be experienced in response to different stimuli, or may be experienced to a different extent, by different individuals. A rabbit's response may depend on the 'personality' (traits formed through genetic influence or early life experience, such as timidity or boldness) or the 'emotions' (response to different motivations, such as frustration or fear) of the individual. It is hard to influence the personality traits of the rabbit – a fearful rabbit may always be more flighty than a bold rabbit in response to an unfamiliar stimulus. However, it is possible to influence the emotional response to a familiar stimulus through behavioural modification.

Personality

The 'personality' of an animal, often linked with its 'temperament', refers to the expression of individual behavioural traits that are stable over

Figure 3.58: A 'bold' rabbit is more likely to show confident, inquisitive behaviour towards an unfamiliar person.

time and context. An animal's personality can also be viewed as the ways that the individual animal differs from the average for its species in its attitudes or motivations, although there may also be 'personality norms' within a species that differ with gender.

The Five Factor Model has been used to assign personality types to animals. These categories are openness to experience, conscientiousness, extraversion, agreeableness and neuroticism (however, the first two of these are harder to assess in animals). In rabbits, speed of learning, memory retention and 'boldness' (i.e. confidence in familiar and unfamiliar situations, Figure 3.58) will all affect its personality.

Physical features of rabbits may also affect these 'personality' traits. Lop-eared rabbits have a reduced ability to perceive sounds due to the ear deformity, which reduces their startle response to loud sounds (and therefore may improve 'boldness', Figure 3.59), but may also affect the trainability of certain behaviours

(Figure 3.60). Larger rabbits have a reputation for being calmer and more interactive: this may be a selected trait, or merely reflect that large rabbits are much less likely to be picked up by their owners. One study showed that, at the same age, juvenile rabbits with a higher body mass spent longer exploring a novel environment and were less anxious when they reached maturity (Rödel and Monclús, 2011).

Boldness is a desirable trait in rabbits, reducing the incidence of fear-related behaviours and increasing the interaction between the owner and rabbit. In various species, many personality traits are heritable to a certain degree – in cats, for example, there is a genetic component to boldness. This has not been demonstrated in rabbits, but selective breeding of rabbits for desirable behavioural traits (rather than just appearance) would likely create a population of rabbits that cope better with life as pets and, in doing so, make better pets for the owners.

Figure 3.59: Lop-eared rabbits have a reduced ability to perceive sounds.

Figure 3.60: Rabbits with one lop ear may struggle to locate sounds (the comparison between both ears is necessary for spatial resolution of sounds).

> *'Personality is to emotion as climate is to weather.'* Revelle and Scherer, 2009

So, personality refers to a pattern of behaviour, cognition and motivations over time and space – how a rabbit is likely to respond in different situations. An emotion is a mixture of different factors at a particular time point – how a rabbit responds to a particular stimulus at a certain time point. An owner can predict how a rabbit will respond based on its personality, but what the owner actually observes at a given moment is an emotion.

Using the term 'emotions' when talking about animals has been controversial: scientists have typically assessed emotions in humans through questions and answers. Emotions were thought to be one aspect that differentiated humans from animals. However, anyone who has spent time with an animal will see that they respond in different ways to different motivations, and they show specific likes or dislikes. When observing their rabbit, owners may describe their rabbit as 'happy', or 'interested', or 'angry' in different situations – these are emotional responses to different motivations. In the brain, there are specific neural circuits for these responses, and these can be triggered naturally or by specific stimulation of parts of the brain.

Understanding these respective emotional responses to motivations, or 'systems', helps to ascertain the cause of a behaviour rather than just the symptom. In cats, seven emotional–motivational systems have been described, and these can equally be applied to rabbits. These systems are:

♦ desire
♦ frustration
♦ fear–anxiety
♦ play
♦ panic–loss
♦ mating
♦ care.

The desire system

The desire system is stimulated when the rabbit experiences internal imbalances (hunger, thirst), external incentive cues (promise of reward) or past learning (previous experiences in similar situations). When stimulated, this system causes rabbits to explore their environment to seek out resources: food, water, shade, shelter, etc. In this system, the seeking motivation is balanced by the requirement for vigilance: if a rabbit is solely motivated to find food and is not alert for danger, then it is much more likely to be eaten.

The desire system is important for learning: animals are motivated to seek resources, and the system causes the animal to learn predictors of reward (that if the rabbit performs a certain behaviour in response to a cue, it will receive a reward).

The frustration system

The frustration system is stimulated if a rabbit's freedom of action or access to resources is limited. When stimulated, this system causes increases in heart rate, blood pressure and muscle blood flow. The behavioural response may vary: the rabbit may escape from the situation, or may show aggressive behaviours towards the source of the frustration.

Allowing an animal to predict and control its environment is extremely important for its welfare. Ongoing frustration is stressful and may result in an increased expression of frustration-related behaviours, or a decrease in responsiveness to its environment (learned helplessness). Predictability has to be appropriately balanced with stimulation: an environment at either end of the spectrum will cause stress.

The fear–anxiety system

This system is stimulated by unpredictable access to essential resources and by threats to resource security. When stimulated, this system causes similar physiological changes to the frustration system, but the behavioural responses are different. If the danger is distant or inescapable, the rabbit will freeze (in humans, this feeling is called anxiety), but if the danger is close and avoidable the rabbit will flee (in humans, this feeling is called fear). If the rabbit cannot perform either of these responses it may use a repulsion response, like aggression, to protect itself. Stimulation of this system is very unpleasant, and if the brain is stimulated in this way in one place, the animal will avoid the place in future.

Some of the fear responses are innate and some are learned. Some rabbits will show fear behaviours when faced with the scent of a ferret, even if they have never seen a ferret. Rabbits can also learn fear behaviours – if a rabbit learns that when it is put in a cage, it is transported to the vet, it will associate the cage with the fear experienced at the veterinary surgery.

Pain is a specific trigger of the fear–anxiety system. It is both a sensation and a motivator. The specific activation of the system is by actual or perceived tissue damage. As discussed earlier, the behavioural responses are different depending on the intensity and duration of the pain, and animals are highly motivated to avoid situations that cause pain.

The play system

While the purpose may not always be immediately apparent, Jaak Panksepp, writing in *Affective Neuroscience* (2004), suggests that 'a basic urge to play exists among the young of most mammalian species'. Play behaviours can be individual or social. Young rabbits (six weeks of age) show chasing behaviour towards other young rabbits, which is thought to be an expression of social play, and the frequency of this behaviour decreases as the rabbits get older.

Play is important for various reasons: the young animals learn about the social structure, develop relationships with other animals and practise courting and mating skills, and it allows other emotional circuits to be exercised in the safety of the home environment.

Prey species also learn how to avoid predators. Perhaps this is why rabbits perform the twisting, jumping behaviour called a 'binky'. These 'functional play behaviours' are much more common in juvenile rabbits but are still 'practised' by older rabbits: a behaviour that is important to keep muscles honed to escape when needed.

The panic–loss system

In humans, stimulation of this system results in feelings of loneliness, grief and separation distress. In rabbits, the system is stimulated by separation of the animal from its companions. The type of attachment and social dependence in rabbits (unlike in dogs) differs between human–rabbit relationships and rabbit–rabbit relationships: hence why interactions with humans cannot be a substitute for interactions with rabbits.

In humans, early childhood loss is a significant risk factor for depression and panic attacks. Rabbits are separated from other rabbits at point of sale (commonly at eight weeks) and deprived of social companionship for the rest of their lives – this may cause permanent emotional changes. Bonded pairs of rabbits should never be separated: separation causes severe distress in both animals (which can cause gut stasis and even death), and may disrupt the bond between them.

The mating system

This system is involved in attraction and selection of sexual partners. Animals are motivated to reproduce. The type of response to that motivation varies between the sexes. Testosterone stimulates male rabbits to search for numerous sexual interactions. Female rabbits are stimulated to show sexually receptive behaviours when they are fertile.

Most rabbits kept in pairs or groups by pet owners are neutered, hence this system is rarely of much consequence in these animals (although, when an animal is neutered, it may still be motivated to seek out others of the opposite sex due to associations with reward, even if the hormonal motivation for the reproductive behaviour is lacking). This system is more commonly a problem when unneutered males are kept singly and show sexual behaviour towards non-rabbit animate or inanimate objects.

The care system

The care system is stimulated in pregnant females shortly before birth and stimulates maternal behaviour towards the offspring. This system is therefore unlikely to be significant in most pet rabbits (as they are neutered) but may be important in unneutered female rabbits undergoing false pregnancies. The care system stimulates territorial aggression around the nest to protect the young rabbits.

What do good welfare and bad welfare look like?

We've so far looked at the various physical factors, the behavioural factors, and the factors that affect how a rabbit 'feels' about the physical and behavioural aspects of its situation. This framework can be used to assess the likely welfare of a rabbit. However, once a rabbit's welfare has been assessed, what should we do with this information? In what ways does the assessment affect the advice given to the owner?

The term 'quality of life' is often used in conjunction with rabbit welfare assessment. The balance between the positive and negative experiences of an animal results in the 'quality of life': a good quality of life is one where the positive experiences outweigh the negative ones (Table 3.3).

Table 3.3 An example quality-of-life scale used when assessing an animal's welfare.

Category	Description
A good life	The balance of salient positive and negative experiences is strongly positive. Achieved by full compliance with best practice advice well above the minimum requirements of codes of practice or welfare
A life worth living	The balance of salient positive and negative experiences is favourable, but less so. Achieved by full compliance with the minimum requirements of code of practice or welfare that include elements which promote some positive experiences
A life that is neither good nor bad	The neutral point where salient positive and negative experiences are equally balanced
A life worth avoiding	The balance of salient positive and negative experiences is unfavourable, but can be remedied rapidly by veterinary treatment or a change in husbandry practices
A life not worth living	The balance of salient positive and negative experiences is strongly negative and cannot be remedied rapidly so that euthanasia is the only humane alternative

A quality of life scale is shown here. The top two categories describe a good quality of life – a 'life worth living' can be transformed into a 'good life' by helping the owner to identify poor welfare in certain areas and make the requisite changes to improve the rabbit's situation. The lowest category is a 'life not worth living', which is seen in rabbits with severe untreatable health problems that cause pain and limit the rabbit's ability to perform rewarding behaviours. In this situation, the welfare problems are unlikely to be reversible, so euthanasia may be the best option. Note that even those rabbits that have 'a life worth avoiding' can be given a 'good life', providing that the owner has the resources and motivation to improve the rabbit's environment.

Owners and referring vets often describe an animal's husbandry as being 'perfect', when what they mean is that the owner is doing everything in their power to do the best for their animals. This does not necessarily mean that the rabbits' welfare is good.

4

How can I change behaviour?

The last sections have addressed how to take a history and examine the problem behaviour, and how to assess the rabbit's welfare to identify why the behaviour is likely to be occurring. We then need to work out how to modify the behaviour. This section covers various tools that you can use.

- Understanding the rabbit's communication in the short and long term
- Changing behaviour through modifying husbandry
- Changing behaviour through modifying interactions

Changing the behaviour of an animal is far easier if you try to work with the motivations of that animal rather than against them. It is therefore important to understand what the rabbit is trying to achieve with the behaviour.

Understand the rabbit's communication

Rabbits communicate with other rabbits in a variety of ways. Some of these methods communicate information about the individual's state over a timescale of seconds to minutes; other methods communicate information about the rabbit's state over longer timescales.

Minute-to-minute communication strategies

Rabbits communicate information about their minute-to-minute physical or mental state primarily through visual postural signals and auditory signals.

Communication through changes in the face or ears

Eyes

A rabbit that is anxious or fearful shows increased tension in the muscles of the face. It's hard to see this through the fur of the face, but you can see that the eyes look bigger. A fearful rabbit's eyes appear to be 'bulging', the sclera (white of the eye) may be visible and they will blink less (Figure 4.1).

A relaxed rabbit may sit with eyes partially closed. A painful rabbit may also have partially closed eyes, so it is important to bring together different pieces of information when diagnosing the emotional state (Figure 4.2).

Lips

A rabbit that is interested or investigating an object has rounded upper lips. If the rabbit is investigating an object on one side of its face, it will flatten the upper lip on that side (Figure 4.3). In a fearful rabbit, both lips are often flattened due to increased muscular tension, which widens the nostrils.

Figure 4.1: A fearful rabbit contracts its facial muscles, so the eyes seem to be 'bulging' from the head.

Figure 4.2: Pain behaviours in rabbits can be very hard to assess. In this image, the grey rabbit on the right has been recently neutered. She is showing partially closed eyes and increased facial tension. Her companion (on the left) is showing a more normal reaction to the stressful situation.

Figure 4.3: Unilaterally flattened lips (when investigating something or looking in one direction) show that the facial muscles are generally relaxed.

Ears

The ears of a rabbit are very important for social communication: perhaps because they are large and mobile (the visual system has evolved to sense movement over a very wide area, and is not well designed for high-resolution sight at close quarters). Ear position is therefore easy for other rabbits to see. This is compromised in rabbits with ear deformities (Figure 4.4).

A rabbit that is investigating its environment will have both ears alert with the aperture (inside of the pinna) facing forwards (Figure 4.5). If there is an unusual or unexpected sound from another direction, or the rabbit is paying attention to a sound while looking in another direction, one ear will be rotated and trained on the source of the sound (Figure 4.6).

A rabbit that is relaxed will typically have its ears in a less upright position than one that is exploring its environment. The inside of the pinnae often face outwards, rather than forwards, and the aperture is not as wide as the muscles are relaxed (Figure 4.7).

A rabbit that is very relaxed will lay its ears along its back. A rabbit that is cold will do the same – the nape of the neck has thinner fur, so body heat can warm the ears in this area.

A rabbit that is scared will flatten its ears against its back, often with widened eyes, a raised tail and a low body position. It may show a 'freeze' response or a 'fight' response. If the fear-inducing stimulus persists, the rabbit is likely to show aggressive behaviour – a lunge, scratch with the forepaws or a bite.

Communication through postural changes of the head

Lowered head

This solicits mutual grooming, and may be either a response to a similar behaviour from

Figure 4.4: Rabbits bred for ear deformities cannot use their ears for communication. This may reduce the ability of their owner, or of companion rabbits, to read their emotional state.

Figure 4.5: A rabbit that is very alert will have both ears focused on the source of interest.

How can I change behaviour?   113

Figure 4.6: Rabbits with normal ears will often train one ear on the source of a novel sound.

Figure 4.7: The ears of a relaxed rabbit (on the left) are less upright and face outwards.

another rabbit (in which the subordinate rabbit typically grooms the dominant rabbit), or may be spontaneously offered. A rabbit showing this behaviour is relaxed and willing to interact.

Head shake or flick

This may be part of the grooming sequence. If seen on its own, a headshake indicates frustration, disinclination to engage in an interaction, or ear discomfort. Also, some rabbits do partial 'binkies', where the paws do not leave the ground, but they show an exaggerated head flick as functional play behaviour.

Communication through postural changes of the body

Hunched

This refers to a small, compact, rounded body shape. The position may be adopted by a rabbit in pain, or by a rabbit that is relaxed: the two must be differentiated by other behaviours and response to normal stimuli (such as food). The likelihood of this position being adopted depends partially on the body biomechanics of the rabbit breed. Small rabbits with short bodies and short legs are more likely to rest with their forepaws tucked underneath them: larger rabbits are more likely to rest in a more flattened position with forepaws extended. Additionally, there is individual variation in resting preference.

Flattened

In this position, the rabbit tries to reduce its visibility to predators by lengthening and flattening its body against the ground and lowering its ears and tail, usually accompanied by widened eyes and increased facial tension. This shows that the rabbit is very fearful.

Sitting

In this position, the rabbit's thorax and abdomen are raised off the ground – the posture is similar to that of a sitting cat or dog. It is a resting position but the rabbit is still alert to changes in the environment.

Lying down

When a rabbit lies down, the thorax is in contact with the ground, and the legs may or may not be outstretched: the degree to which the legs are outstretched signifies the degree of relaxation. This behaviour is also affected by temperature: in warmer conditions, rabbits may stretch out in the shade to increase heat transfer away from the body, and in cooler conditions, rabbits avoid stretching out because this increases their surface area from which to lose heat. Once again, this behaviour may not always indicate relaxation: it is also shown in very stressed rabbits (Figure 4.8).

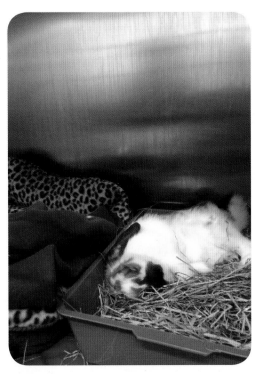

Figure 4.8: Some rabbits show postures that seem very relaxed in stressful situations (such as in the veterinary hospital). Context is very important when assessing a rabbit's emotional state.

Tail position

Like the ears, the tail protrusiong is visible to other rabbits. The tail can be protruded away from the body: this typically occurs in several situations. A rabbit that is fleeing from danger, or in preparation to run, will do this to make the scut (underside of the tail, white in wild rabbits) more visible to its conspecifics. Rabbits also protrude the tail when in states of high arousal: when showing agonistic behaviours or sexual behaviours.

Communication through gait
Binkying

This term describes a set of behaviours that have no obvious function in isolation: either a 'full binky', where the rabbit jumps into the air and flicks its hind paws, or a 'partial binky', where the rabbit shakes its head, with or without a jump. These behaviours are thought to have a role in escape training – it is useful for the rabbit to 'practise' high speed running and changes of direction so it is able to escape from predators when it needs to. These behaviours are frequently performed when the rabbit is relaxed and in an environment where it feels safe, and are performed more when the rabbit is moved from a confined environment into a larger safe environment.

Sprinting

Short sprints may be performed with or without a binky. If not triggered by an external stimulus, this seems to be a functional 'play behaviour', and is usually performed under similar conditions to binkying.

Chasing

Young rabbits chase and displace each other as a form of play behaviour, which helps to establish social hierarchies at a young age. Older rabbits chase each other more rarely. This behaviour may occur when there is some competition over a resource: when one rabbit has a piece of food, for example, or when one rabbit is trying to mount the other (which can be shown by females or males, neutered or unneutered). The chasing rabbit often has a raised tail and each hop may be higher than normal. This behaviour is a common way of resolving conflicts: the chase usually stops when the chased rabbit turns towards the chasing rabbit, lowers the head and flattens the ears.

Flicking hind paws

This behaviour may be shown if a rabbit is withdrawing slowly (i.e. not fleeing) after frustration or irritation. The hind paws are flicked with each step, which may make a sound. The behaviour is also shown if a rabbit has wet hind paws, as it is an effective way of removing excess water.

Communication through sound
Teeth grinding

A rabbit often chatters its teeth quietly when it is relaxed, calm and finds a situation pleasurable. This can be when groomed by a companion or human, and sometimes as the rabbit settles after grooming itself. As with posture, this behaviour is context specific – rabbits may grind their teeth if acutely painful. However, this grinding is often louder.

Grunt

The grunt, a quiet, low-pitched, guttural sound, is commonly associated with sexual behaviours from male rabbits. Rabbits may also make this sound if they are in pain, or if they are unexpectedly picked up.

Growl

A rabbit growl is of short duration (reflecting the small size of the thorax), and is the auditory component of agonistic behaviour: a rabbit that growls may show flattened ears and widened eyes. Rabbits may also growl in confrontation with other rabbits over food resources. A growl indicates a high level of arousal and stress. In rare cases, rabbits may growl in frustration.

Thump

Rabbits thump their hind paws on the ground to produce a longer-distance auditory warning of danger to others in the group. Thumping is often a social communication of fear or anxiety, but can also indicate frustration. Occasionally it is used in conjunction with functional play behaviours, such as binkying or short sprints.

Scream

Rabbits can produce a high-pitched scream if extremely fearful or in acute pain. This behaviour is thought to have evolved to distract a predator and allow the rabbit to make an escape.

Longer-term communication strategies

Rabbits communicate information about their longer-term physical or mental state primarily through olfactory signals from skin scent glands, although these olfactory signals may be marked by visual cues. Rabbits may transmit these signals actively (by chin-marking, urine spraying or defecation) or passively (mammary gland and inguinal gland secretions). Scent-based communication in rabbits is far more complex than humans can detect. The rabbit has an organ in its nose called the vomeronasal organ, which detects pheromones. The principal sources of odour that inform social behaviour are from the submandibular glands, urine and the anal gland (Melo and González-Mariscal, 2010). Diagram 4.1 shows the distribution of key scent glands around the body.

Communication through submandibular pheromones

The scent gland under the rabbit's chin allows scent marking when the rabbit performs the behaviour called 'chinning'. The rabbit rubs its chin on different objects: grass blades, stones, entrance of a burrow, a subordinate rabbit or on dunghills. In the behaviour, the rabbit puts its jaw on the object to be marked and pushes

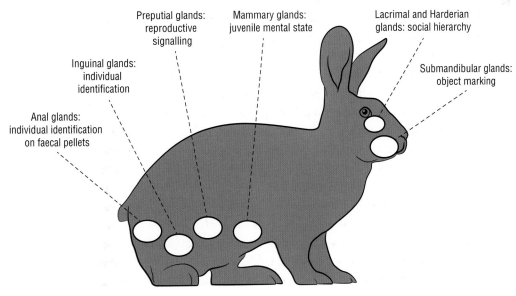

Preputial glands:
reproductive
signalling

Mammary glands:
juvenile mental state

Lacrimal and Harderian
glands: social hierarchy

Inguinal glands:
individual
identification

Submandibular glands:
object marking

Anal glands:
individual identification
on faecal pellets

Diagram 4.1: Location of rabbit scent glands.

Figure 4.9: Rabbit 'chinning', marking an object with secretions from its submandibular glands.

the head forward, leaving behind submandibular gland secretions (Figure 4.9).

Unmated female rabbits (does) show 'peaks' of chinning every four to six days, and, within an hour after mating, there is a dramatic decrease in this behaviour (González-Mariscal et al., 1990). Pregnant does continue to show a low level of chinning, and the behaviour slowly increases again after parturition. The glands are larger in males, and are reduced when the male animal is neutered. In both domestic and wild rabbits, dominant rabbits chin-mark more than subordinate rabbits. Rabbits are more likely to chin an object if another rabbit has previously marked it. Rabbits may chin their owner's skin, but this is uncommon – the scent marks are usually left on static objects. This rarely causes a problem for owners because the submandibular gland secretions are imperceptible.

This behaviour is thought to have several functions: to establish and maintain social rank, to define territory and to enhance self-confidence. Chinning is more common within a rabbit's own territory. The pheromones also provide information on the social and sexual status of other individuals in the area.

Communication through urine marking

Rabbits also use urine as an olfactory signal. When used as a signal, it is deposited by 'spraying': the rabbit runs past the target (an object, another rabbit or another animal) at speed, rotates its hind legs and sprays a jet of urine. This is a short-term communication strategy if sprayed on another rabbit (enurination), and longer term if an object is marked. Dominant males perform this behaviour more frequently than other rabbits. Mature females spend longer investigating urine from dominant males than they do from subordinate males: suggesting that the urine transmits information about a male's testosterone levels and

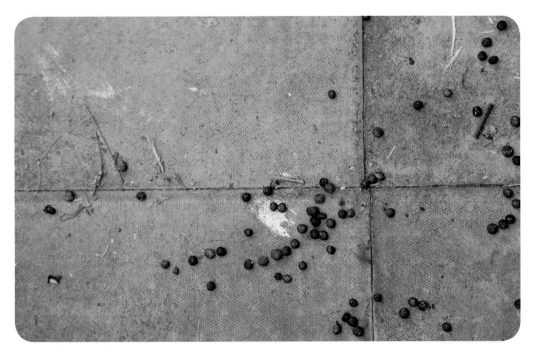

Figure 4.10: Most rabbits will deposit some faecal pellets on the floor, away from the latrine.

fertility. Neutering a male reduces the incidence of urine spraying.

Urine may also be used to mark territory. The scent of a strange rabbit's urine may increase the incidence of aggressive behaviours. Urine deposited as part of normal elimination also appears to have a marking function: rabbits will often urinate where another rabbit has urinated. This non-sprayed urine marking behaviour does not seem to be as affected by neutering.

Communication through anal gland pheromones

Anal glands are found in both female and male rabbits, and appear to be under voluntary control. The rabbit can compress the gland to change the olfactory signal deposited on the faecal pellets. The odour is stronger with increasing age of the rabbit, is stronger in males than in females and is maximal during the breeding season. Pellets deposited for marking purposes smell more strongly (the odour is,

unsurprisingly, described as 'rabbity'), and these pellets are often deposited outside the home cage (Schalken 1976).

Rabbits defecate in many places around their home range, but they also deposit pellets more in certain sites (Sneddon, 1991), called 'dunghills' or 'latrines' (or litter trays, for house rabbits). All rabbits defecate in these sites more frequently than they do elsewhere (dominant male rabbits visit these sites more frequently than other animals) (Figure 4.10). Rabbits often spend time at the latrines sitting and lying down – this may be used to increase transfer of the scents to the body of the rabbit.

Communication through mammary pheromones

Mammary pheromones are important in the young rabbit as they influence learning and fear responses. Does release a pheromone from around the nipple, the mammary pheromone, which causes changes in behaviour in the newborn rabbit kit. It helps the kit to locate the nipple quickly, it promotes learning of new scents and it helps the kit to learn about mixtures of scents, which are important for pattern recognition of 'good' and 'bad' scents later in life (Coureaud et al., 2010). Exposure of neonatal rabbits to human scent around the time of nursing has been shown to decrease their fear response to humans later in life. This is extremely important for reducing stress responses in adult rabbits.

This scent is not used to communicate with owners. However, the similar pheromone in dogs has been synthesised and is effective at reducing stress-related behaviours. If behavioural problems in rabbits are ever seen as a lucrative market, then it is quite possible that this rabbit pheromone may have commercial use.

Communication through inguinal gland pheromones

Inguinal gland pheromones are important for individual rabbit identification. When the offspring of female rabbits were rubbed with inguinal secretions of other females, the mothers sniffed and nudged them more, and even chased and bit them. A rabbit that has been smeared with the inguinal gland secretions (as compared to urine or submandibular gland secretions) of an unfamiliar rabbit will be attacked by its companions.

Communication through ocular pheromones

Ocular pheromones are released from the lacrimal gland and Harderian gland, which are located around the eyeball. The glands are larger and more active in dominant rabbits, so seem to play a role in social hierarchy. These glands not only release pheromones, but also secrete lipid, which creates a very stable tear film, allowing rabbits to blink much less frequently than humans.

Communication through preputial gland pheromones

The preputial glands are found in both females and males, and are located close to the penis or the vagina. In females, these are thought to add pheromones to the urine as part of reproductive advertisement and marking behaviour. Their role in males is still uncertain, but is likely to play a role in reproductive behaviour.

Changing behaviour through changing environment

When we understand what a rabbit is trying to communicate, we can see how it perceives its situation. It allows both an initial diagnosis, and a subsequent assessment of the effect of any changes made.

Let's start by thinking about how to change behaviour by changing the environment. Understanding the physical factors that contribute to an animal's perception of its situation is key to identifying the areas that should be changed. In many situations, merely resolving the welfare problems is sufficient to prevent the unwanted behaviours.

Comparing the effect of diet, environment and companionship on behaviour

In previous sections, we explored how different husbandry practices affect rabbit behaviour. For better comparison, the graphs of each study are replicated in Diagram 4.2.

As shown, the single intervention that is most effective at improving a rabbit's welfare is providing another rabbit. Companionship gives numerous positive emotional benefits: it is the most interactive, stimulating form of environmental enrichment, it reduces the incidence of stress-related disease and it increases

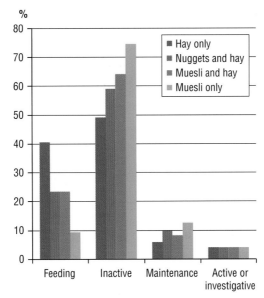

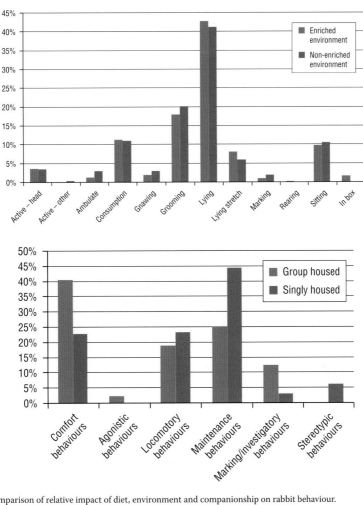

Diagram 4.2: Comparison of relative impact of diet, environment and companionship on rabbit behaviour.

the rabbit's willingness to spend time eating hay or grass. Companionship affects all of the physical domains that contribute to a rabbit's mental state.

In the following section, we'll explore how to change the environment to improve the rabbit's welfare and resolve unwanted behaviours.

How to provide companionship

Rabbits require the companionship of other rabbits for a good life. While companionship may be critical for good welfare, in most cases, a rabbit does not initially see an unfamiliar rabbit as a prospective companion. Rabbits need to be 'bonded' together – a process of meeting, negotiating terms and accepting a contract! A 'bonded' pair of rabbits shows much affiliative behaviour and little aggressive behaviour towards each other.

Stable rabbit pairings

Some pairs of rabbits are more likely to bond successfully than others. The most stable pairing is a male–female pairing: this replicates the situation in the wild, providing that the male rabbit, at least, is neutered. This pairing is usually stable throughout the life of the rabbits.

With male–male or female–female pairings, the likelihood of success depends on the age of the rabbits. Keeping a pair of littermates together, or introducing baby rabbits to each other at a young age, can lead to a stable bond, but this is much more likely for a female–female pair than a male–male pair. Frequently, male–male pairs fight viciously and unexpectedly when they reach puberty and can cause significant injury to each other. If a fight has happened, the chance of successfully re-bonding the rabbits, even if both are neutered, is very low. If such a fight has occurred, owners should rehome one of the

male rabbits and bond the other to a suitable female. If owners wish to acquire a single-sex pair of rabbits as juveniles, then they are likely to have more success with a pair of females than a pair of males (Figure 4.11).

Paradoxically, if the rabbits are being introduced as adults, then some rescue centres find that neutered male pairs (providing there has been no history of a fight) can be easier to bond than neutered female pairs, as the male rabbits tend to show less territorial behaviour.

Although many single-sex pairs of rabbits do develop a strong bond, the easiest pairing to bond is a male rabbit and a female rabbit. However, there are a few challenges to keeping rabbits in this way. First, rabbits reproduce prolifically: if large numbers of offspring are not desired, then the rabbits need to be neutered. Second, rabbits are very territorial, so introductions need to be performed carefully to prevent injury to either animal.

Let's address these challenges in turn.

Preventing reproduction

Neutering rabbits kept in pairs or groups is vital to prevent reproduction. Removing the ovaries in females reduces the risk of uterine and mammary tumours and reduces the risk of territorial aggression. Removing the testes in males reduces urine spraying and undesired sexual behaviours. Behaviourally, testosterone is important for confidence in males, so when this is removed the rabbit may seem more fearful – perhaps both because of the reduction in testosterone and the aversive nature of the procedure. Some owners report that the rabbit's personality changes slightly in the short to long term after neutering. Owners should also be aware that, even after neutering, two rabbits are likely to mount each other occasionally, especially during the spring. Some hormones regulating sexual behaviour

Figure 4.11: Juvenile rabbits are unlikely to fight when introduced, but the bond may break down at puberty if the rabbits are not neutered.

are released by the brain, not the reproductive organs.

While neutering is essential for rabbits kept with other rabbits (the welfare gain of species-specific companionship outweighs the welfare cost of the procedure), the procedure itself causes a lot of pain and distress, and 1 in 150 rabbits die under anaesthetic. Diagram 4.3 shows a decision-making process for determining whether or not to neuter a rabbit.

Case study 4: Fudge

Fudge was a six-month-old female dwarf rabbit. She had been bought as a pet for the owner's children, and was housed alone in a hutch without access to an exercise space. She was regularly handled by both the owner and the children and was moved to a run for exercise.

Fudge was booked in for a routine spay as advised by the veterinary surgeon. Three days after her surgery, at the postoperative check, the owner reported that she did not seem painful but had started showing aggressive behaviour towards the owner's father, biting him when he attempted to lift her. The vet gave more pain relief.

A week after surgery, she had started to show aggressive behaviour to the whole family and they were no longer able to handle Fudge or move her to her run. The owner was not willing to try to desensitise and counter-condition their rabbit's response to people, preferring to wait and see if the problem resolved. At subsequent follow-up, the rabbit's behaviour had not improved and the owner was trying to rehome her.

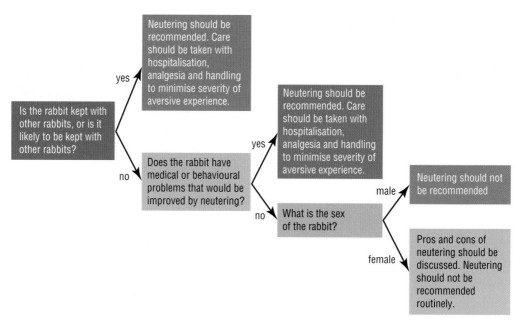

Diagram 4.3: Decision tree for deciding whether to neuter a rabbit.

Neutering causes both pain and distress. Rabbits experience pain because their internal organs are manipulated and slightly damaged during the surgery (especially in female rabbits), because of the skin incision and tension on the stitches during recovery and because large volumes of drugs are injected into the muscle. Pain is hard to assess in rabbits, as previously discussed, and so frequently is not adequately treated. Rabbits experience significant stress when they are transported, put in a kennel near unfamiliar predator species, when they are restrained, when they are handled to check the wound and when they are transported back to the vet for postoperative visits. Additionally, the fear responses of the rabbit are suppressed when they are juveniles, but are no longer suppressed when they go through puberty: so the timing of neutering means that the experience is worse (Bradbury and Dickens, 2016).

A very aversive event, such as being neutered, may also erode the rabbit's trust of humans and worsen relations between owner and pet. This is more problematic if a rabbit is kept alone, since single rabbits are more dependent on the owner for social interactions. One study found that the instance of owner-directed aggression was more prevalent in neutered versus intact males (d'Ovidio et al., 2016) (the study could not ascertain if this was because of the neutering procedure or hormonal effects, or whether the rabbits had been neutered because they were aggressive).

For pairs of rabbits, the welfare gains from being neutered outweigh the welfare costs, and being transported and housed with a bonded companion will lessen the stress of the procedure. For single rabbits, this stress can, of course, be minimised by good veterinary care and adequate pain management. However, even if all veterinary practices developed the facilities and knowledge to provide for these needs, the stress of the procedure could not be completely removed. This highly stressful experience may worsen the bond between rabbit and owner, and may explain the anecdotal reports of 'change in personality' in neutered rabbits.

To reiterate, if an owner does not plan on keeping the rabbit with another rabbit, and there are no sex-hormone-related behavioural problems, neutering should not be advised. The procedure causes such pain and suffering that, in singly housed rabbits, the welfare cost of neutering outweighs the benefits.

This has not always been thought to be the case, but recent evidence shows that the health benefits of the procedure may be less significant than previously believed (most uterine and mammary tumours occur in rabbits at an age that is well over that reached by most singly housed rabbits [Whitehead, 2015; Whitehead, 2017]).

Additionally, it is difficult to justify performing a painful procedure to extend the life of a rabbit that will be kept under conditions that cause very poor welfare.

There has been additional discussion over whether it is appropriate to neuter female rabbits at all. A proportion of unneutered female rabbits do develop uterine tumours later in life, but these often do not cause clinical signs and are not life limiting – additionally, the average age at which the tumours are diagnosed (often incidentally) is greater than the average age that pet rabbits reach. Also, abdominal surgery in rabbits frequently causes adhesions (where the internal organs become stuck together): this can cause chronic pain, sometimes even necessitating euthanasia. In a male–female pair, the male should always be neutered (to prevent reproduction), but it may be appropriate to leave the female unneutered unless there are specific medical or behavioural justifications for the procedure. There is currently insufficient research to support an evidence-based approach to neutering female rabbits.

If rabbits are neutered just prior to being introduced, advise the owner to wait about six weeks before attempting introductions. This gives enough time for the male rabbit's hormone levels to reduce and enough time for the incision to heal.

In general, neutering rabbits should only be performed if the rabbit is going to be kept with another rabbit, or if there is a behavioural problem in a singly housed rabbit that can only be resolved by neutering it.

Preventing territorial aggression

The paradox when introducing or 'bonding' two unfamiliar rabbits is that both rabbits will have better welfare once familiar, but their instinctive dislike of unfamiliar rabbits on their own territory can make the introduction process very stressful.

There are three ways in which owners can achieve a bonded pair of rabbits.

- Acquire two new rabbits at the same time.
- Bond their current rabbit to another rabbit at a rescue centre.
- Bond their current rabbit to another rabbit at home.

Let's explore these options.

Acquire two new rabbits at the same time

The easiest way to end up with a pair of bonded rabbits is to obtain a pair of bonded rabbits. Any owner who seeks advice on obtaining a rabbit should be advised to get two together (and should be warned about getting male–male pairings, which rarely work). Many pet shops now give discounts to incentivise purchase of two rabbits rather than one ('one rabbit for £30, two rabbits for £40') and the staff are often trained to promote this. If a male–female pair of rabbits is bought at eight weeks, the owner will need to neuter one of the rabbits before either rabbit reaches puberty: the RWAF recommends neutering male rabbits before 16 weeks, and female rabbits at around 20 weeks. Both rabbits should be taken to the veterinary surgery each time, to prevent disruption to the bond and to reduce the stress of the hospitalisation (Figure 4.12).

If owners have bought two males, they are likely to form a strong bond until puberty, when they often start to fight. Neutering is not normally effective at resolving this, but very early neutering may reduce the risk of fights and bond breakdown. If an owner reports that their veterinary surgeon will not neuter their pair of male rabbits as soon as the testicles have descended, then the owner should be advised to find a veterinary surgeon that will. Some vets prefer to perform the surgery when the rabbits are six months old, but this is likely to cause permanent bond breakdown.

Good rescue centres usually have bonded pairs that can be rehomed; if they don't, they are usually happy to bond two of their single rabbits.

Some owners may purchase two rabbits that have not been previously kept together and introduce them at the same time to a novel environment. The risk of territorial behaviour is lower with this approach than if one rabbit is introduced to another rabbit's territory (especially if both rabbits are juvenile), but owners should be alert for aggressive behaviour between the rabbits: see next section.

Figure 4.12: If one of a pair of bonded rabbits requires veterinary treatment, a good veterinary practice will encourage hospitalisation of both rabbits. In this practice, hospitalised rabbits are kept in the isolation area to reduce sight, sound and scent contact with predators.

Bond their current rabbit to another rabbit at a rescue centre

Most rescue centres that rehome rabbits will have the facilities to offer owners a bonding service. This increases the likelihood of a successful pairing (in the best interests of the owners, the rabbits and the rescue centre).

Rescue centres are usually more successful at bonding rabbits than owners are. There are several reasons for this.

- Rabbits at rescue centres are usually more stressed (recently arrived rabbits have not acclimatised to the new environment, there is more exposure to different people and more scents of other rabbits) so are more likely to see another rabbit as an ally rather than a threat.
- Rescue centres have environments that are completely unfamiliar to both rabbits, and that strongly smell of other rabbits, so territorial aggression is less common.
- Rescue centre staff have much more experience of 'normal' and 'abnormal' behaviour when introducing new rabbits, so are more likely to intervene only when necessary.
- Many rescue centres also ask prospective owners to list their 'top three' rabbits. Sometimes, certain rabbit–rabbit pairings are not successful, so being able to try the rabbit with other rabbits will increase the likelihood of a successful bond.

Typically, prospective owners will bring their current rabbit to the rescue centre. The environment will be set up without resource focuses (no food bowl, no hutch to guard, hay scattered on the floor) and with distraction (concentrate food scattered on the floor). The rabbits will be placed into the room and observed continuously for the first hour or so, and then

Figure 4.13: Rabbits bonding at a rescue centre. The rabbits have been together for four days and are now resting in close proximity, which indicates that the bond is established.

intermittently over the next few days. Bonding typically takes from three to seven days, and the rabbits are ready to go back to the owner's house when they are comfortably sitting in relatively close proximity (Figure 4.13).

The owner should transport the rabbits back to the house in one box. The home environment should have been disinfected and scrubbed to remove scent traces of the first rabbit, and, where possible, the environment should look different (cardboard boxes can be arranged to change the appearance).

Some degree of chasing and mounting behaviour is common over the first week or so. This can be concerning to the owners, as male rabbits frequently pull out hair from the middle of the back of the female rabbits. Owners should provide distraction and avoid intervening, providing that the rabbits are not showing overt aggression (see section on Aggression at p. 173). The mounting behaviour will reduce in frequency, but you may need to provide reassurance to the owners at this time. Punishing one rabbit, or separating them for periods, will slow the formation of the bond and is unlikely to prevent the mounting behaviour.

Bond their current rabbit to another rabbit at home

The principles of bonding rabbits at home are the same as bonding rabbits in a rescue centre; however, the pre-existing territory of one rabbit can increase the risk of territorial aggression. There are two approaches to this: either to create a novel territory on which to introduce the rabbits (fast-bonding approach), or to increase familiarity between the rabbits without giving them opportunity to injure each other (slow-bonding approach). The success rate seems to be better if a female rabbit is introduced to a resident male rabbit than vice versa.

There are advantages and disadvantages of both of these approaches. The former can be referred to as a 'fast-bonding' approach: the method creates an unfamiliar territory into which the rabbits are introduced immediately and left to establish a relationship. The major advantage of this approach is that if a bond is successful, it is usually fairly quick to establish (between three and seven days). The disadvantages are that the success rate may be lower: it is difficult to create a novel territory in an environment that is familiar to one rabbit, and owners who are not familiar with the bonding process may be too quick to separate rabbits that are establishing a hierarchy.

Creating a truly novel territory can be hard: although the scent of the rabbit can be removed with enzymatic cleaner, there are other scents in the house (of the owner, common cleaning products, fragrances, etc.) that comprise the scent-familiarity of a territory. Additionally, creating a novel visual environment can be difficult, and require significant reorganisation (and may not be successful). If the owner has an alternative space (room, shed or garage) with which neither rabbit is familiar, they may have more success.

For the second approach, the 'slow-bonding' approach, the rabbits are put in neighbouring enclosures where they can have visual, olfactory and some physical contact through the mesh for the rest of the time (to increase familiarity, Figure 4.14). The advantage of this approach is that the risk of injury is lower (very important if the rabbits have shown aggression to each other), but the disadvantages are that this process takes a long time, and the owner needs to be able to divide the environment into two adjacent enclosures separated by mesh. Preferably, initially there should be a gap between the enclosures so the rabbits cannot actually reach each other (to prevent them injuring each other through the mesh). The gap should be reduced as the rabbits become more familiar.

When the rabbits are interacting well through the mesh, the mesh can be removed for short periods to assess how they interact. As with rescue-centre bonding, some amount of mounting behaviour is common but this will gradually diminish in frequency.

Older sources of information advise putting unfamiliar rabbits together in stressful situations to bond them together, such as in the car or in the bath. However, a bond formed this way is likely to be weak and easily disrupted when the rabbits are in a non-stressful environment. Additionally, by definition, it causes a lot of stress to the rabbits. Therefore, you should not recommend it to owners.

Behavioural responses during bonding processes

There are a variety of behavioural responses that may be seen when rabbits are introduced (aggressive behaviours are described in more detail in Section on Aggression at p. 163).

Rabbits ignore each other

This is also a good sign, as the familiarity between the rabbits will gradually increase without agonistic behaviour.

Figure 4.14: When slow bonding rabbits, they should be kept in neighbouring enclosures to allow visual, olfactory and some physical contact.

Rabbits groom each other

This is a very good sign, but there may be a second period of disruption before a solid bond forms. Rabbits that have been kept singly may show aberrant responses to normal social behaviour: such as soliciting grooming by lowering the head, but then responding with surprise and alarm to being groomed and nipping the other rabbit. Inexperienced rabbits may try to groom the other rabbit's paws or flanks. This behaviour will decrease with time as the rabbit learns appropriate social interactions.

One rabbit jumps over the other rabbit

Both rabbits often show this behaviour. The jumping rabbit may strike the back of the other rabbit, dislodging clumps of fur. This looks dramatic but rarely causes injury; it is part of the process of establishing a hierarchy, and shouldn't be prevented.

One rabbit mounts the other (usually male mounting female)

If the mounted rabbit submits to this, it is not a problem (males often pull fur out of the middle of the back, which can look dramatic, but providing the female doesn't object, this is not a problem). If she does object and runs away (and has space to do this), this is not a problem. If she becomes aggressive, then the introduction will need to be slow.

One rabbit interacts, one rabbit doesn't

The first rabbit may interact in an affiliative or agonistic way and the other rabbit flattens itself and freezes. This doesn't usually cause a problem: submissive behaviours rarely trigger overt aggression. However, if the interacting rabbit's behaviour is overzealous (often grooming or other attentive behaviours), the submissive rabbit may eventually become frustrated and show aggressive behaviour back. If this behaviour is likely to cause a problem, it may be worth separating the rabbits before they fight and moving to a slow-bonding approach.

One rabbit chases, one rabbit runs away

This is not a problem providing that the chased rabbit does not fight back and does not get hurt. However, if injury looks likely, the rabbits should be separated and the bonding process should be slowed down.

Rabbits fight

The rabbits should be separated and the bonding process should be slow. Rabbits should not be allowed to 'fight it out' as the behaviour is more likely to be repeated, and the risk of serious injury is very high.

It can be really difficult to bond some rabbits (often female) with another rabbit. Typically, they groom the face of the other rabbit and then groom the flanks or underside, often nipping the rabbit in the process. This behaviour is likely to reduce in frequency as the pair of rabbits reaches equilibrium, but may cause aggressive behaviour from the other rabbit, in which case the bonding process has to be stopped and another partner tried. If the other rabbit has plenty of space to move away, this may not cause a fight: owners should be reassured that the behaviour will decline in frequency over time, and providing one rabbit defers to the other, it is a normal part of the bonding process.

Such rabbits may benefit from the slow-bonding approach. In this situation, owners are advised to seek help from an experienced rescue centre. Good rescue centres may either be willing to try to bond the rabbits themselves, or should be willing to allow an experienced owner to try a slow-bonding process at home (and therefore be willing to take the new rabbit back if the process is unsuccessful).

Providing a more stimulating environment

As previously discussed, providing a companion yields more benefits to the rabbit than enriching its environment – getting a second rabbit is a better welfare investment than improving the environment. However, improving the environment will allow the rabbit to express more normal behaviours and experience more positive affects.

Rabbits are very dependent on their environment and territory for emotional stability, so any changes should be gradual. If the rabbit is kept singly, this is even more important, as it lacks the continuity of a companion to cope with these changes. If a rabbit's environment is to be enlarged, the rabbit should be given the choice about how and when it ventures into the new space. Owners can train a recall command and use that to reduce fear, or they can scatter food on the ground of the new area, but picking the rabbit up and putting it into unfamiliar territory can be stressful. Rabbits that have never had access to large spaces may seem agoraphobic, and it may be some weeks before they gain confidence to explore.

One barrier to owners providing a more stimulating environment for their rabbits is their perception of relative risk. Owners are sometimes rightly concerned about the risk of their rabbits being injured or killed by roaming predatory pets or wildlife: the risk of physical suffering is more conspicuous than the risk of mental suffering from a restricted environment. However, an imbalanced judgement of risk may mean that the rabbits are unnecessarily restricted. Foxes definitely pose a risk to rabbits: rabbits should not be allowed out unsupervised, especially at night, when foxes are in the neighbourhood. Most cats will not take on large rabbits, so the risk from these is lower. If the rabbits can be let out in a garden, there should be multiple places to hide if the rabbits feel threatened, and the rabbits should have easy access to an area that is safe from predators (if a cat flap is available, the house can be a good refuge).

Advice sheets 5 and 6 in the Appendix are for owners wishing to improve their rabbit's environment, both for house rabbits and for outside rabbits.

Reducing the need to pick up a rabbit

We have already discussed how much rabbits dislike being picked up. This unpleasant interaction is one of the most common root causes of fearful and aggressive behaviour towards the owner. If the owner relies on picking up the rabbits for practical reasons, they may find it very hard to see how they can avoid this, so a two-way discussion is essential.

Owners tend to lift rabbits for certain purposes:

- Remove them from their cage in order to interact with them
- Move them from their hutch to their run
- Perform a health examination

We'll now explore how owners can achieve the desired outcome without picking up their rabbits.

Remove them from their cage in order to interact with them

In order to get a rabbit out of its cage without picking it up, the owner needs to teach the rabbit that coming out of its cage is rewarding. They can do this in a conscious way (by training the rabbit), or in an unconscious way (by placing food or novel objects outside the cage).

It is really important that owners avoid picking up rabbits when they want to show affection to them. If a rabbit bites an owner when they pick it up, the owner feels that they have been punished for showing affection to the rabbit. Even if the owner successfully lifts the rabbit out of the hutch, the rabbit will be in a higher arousal state and less inclined to act in a calm, affiliative way towards the owner.

A better solution is to train the rabbit to come over on command (see section on Training). The hutch or cage should be set up in such a way that the rabbit can come out voluntarily to interact with the owner. Owners can provide a shallow-gradient ramp for high hutches and put a door in the mesh of ground-level runs.

Owners should understand that the process of building trust is a slow one, and so it will take a while for the rabbit to build new, positive memories about the owner to outweigh the negative associations with being picked up. Setting reasonable expectations is vital to ensure that the owner does not become disillusioned and revert to picking the rabbit up to interact with it.

Move them from their hutch to their run

If the hutch is always connected directly to the run, the owner will not need to pick up their rabbits to move them. The RWAF recommends that outdoor rabbits have continuous access to a run, and that house rabbits have continuous access to a room. However, many hutches for sale are not easily connected to a run. If the hutch and run are separate, and the rabbit becomes fearful or aggressive when picked up, the effect on their welfare will be not only through a worsened relationship with their owner, but also through lack of space. If owners can't afford, or don't want to buy, a more appropriate hutch, there are ways to modify existing hutches: using mesh tubes, attaching ramps or providing a step. There is plenty of information on connecting parts of the rabbits' environment on the RWAF website.

Recall training will also allow an owner to move the rabbits around without lifting them up. If an owner has an enclosed garden, rabbits can be easily trained to use a cat flap or pet door. If the rabbits are microchipped, a cat flap that opens on microchip recognition will prevent any neighbourhood cats following them into the house (a low risk, but possible). If the environment is safe, the rabbits may be allowed free access to the garden. Alternatively, the owner can unlock the cat flap when they are at home, to allow the rabbits to use the garden, and then lock it again when they have to leave the rabbits unsupervised.

Perform a health examination

An owner can assess the health of their rabbit in ways that don't make it stressed. Vets routinely recommend that owners assess the rabbit's behaviour daily for signs of illness, and perform a more thorough check-up once a week. Vets and owners often believe that they need to lift the rabbit up and turn it over to assess its health. However, this causes a lot of stress, and if owners are doing it every day, they will have a very poor relationship with their rabbit.

As described in the section on 'Health', providing that the rabbit is trusting of, and comfortable with, the owner, a good clinical examination can be performed without requiring the rabbit to be picked up. When an owner has decided to stop picking up their rabbit, they need to build sufficient trust (through numerous positive interactions without aversive consequences) for the rabbit to tolerate a thorough examination. During this period, the rabbit must not be picked up, as it will set back the process: the owner must be especially alert to behavioural signs of ill health.

Changing diet

The rabbit's taste preferences and natural diet were discussed in section on Types of rabbit feedstuffs at p. 24. Often, pet rabbits are provided with a diet that doesn't meet their requirements. If this is the case, you will have to advise the owners on how to transition rabbits onto a better diet (which contains more forage and less concentrates and vegetables). Abruptly changing a rabbit's diet can cause severe illness, as the rabbit may stop eating, which causes ileus. Therefore, owners need to proceed carefully. From an evolutionary standpoint, rabbits are evolved to eat fresh plant material, mostly grass and grassland plants, and some leaves and bark. From a behavioural standpoint, rabbit behaviour is adversely affected if they are deprived of access to grass or hay.

Feeding rabbits muesli or concentrate food freely or 'ad libitum' leads to obesity, urinary tract disease, gastrointestinal tract disease and dental disease. Feeding large quantities of high-sugar fruits or vegetables will also often cause health problems.

However, rabbits like to eat these foods, and many owners like to feed rabbits these foods. Therefore, in order to encourage rabbits to eat sufficient grass or hay, it is necessary to restrict, or even eliminate, concentrate food and high-sugar or high-starch vegetables. Owners should be informed that they should offer fresh grass and hay every day, and the rabbits should never 'run out' of these foodstuffs.

What is the effect on a rabbit's welfare of limiting access to highly palatable foods to preserve long-term health? Animal welfare science discusses two forms of food restriction: quantitative food restriction (reducing the total amount of food fed) and qualitative food restriction (reducing the energy density of the food that is fed).

The commonly fed, high-energy diet (ad lib concentrate food) is an unnatural diet for a rabbit: the natural diet is much less energy dense. However, when an owner moves their rabbits from a diet with ad lib concentrate food on to a hay-based diet, this represents some degree of qualitative food restriction. Owners may worry about adverse effects on the rabbits' welfare: that they are 'depriving' or 'starving' the rabbits of 'proper food'.

When animals are restricted in the amount of food that they consume (quantitative restriction), they often spend more time foraging for food and may also spend time on other oral-related behaviours (such as grooming themselves or chewing inedible objects) or stereotypies. The evidence suggests that when animals (pigs and chickens) are fed a diet that is less energy dense, the animals eat more slowly (so spend longer eating). They usually show a normal pattern of meals and intervals between meals, and are less likely to perform other oral-related behaviours (D'Eath, R.B., et al. 2009). This indicates that, where the total calorie consumption needs to be reduced, owners should replace high-calorie foods (concentrates) with low-calorie foods (forage), rather than just restricting the amount of high-calorie food that they feed.

For rabbits that are used to eating a lot of concentrate foods, this transition is likely to cause some degree of transient stress (bad for short-term welfare). However, as previously discussed, it is extremely important for the rabbit's physical and behavioural health, and will greatly improve welfare in the medium and long term.

Ideally, no rabbit would be fed on ad lib concentrate food, so the transition to a forage-based diet would not arise. However, when necessary, owners can reduce the stress caused by this transition in several ways. Reducing the quantity of concentrate food gradually allows a steady increase in quantity of forage consumed. Abruptly stopping feeding concentrate food can cause anorexia and gut stasis, so owners

should never do this. Providing more palatable food as an alternative to concentrate food can also reduce the transition stress: as an interim measure, the owners may slightly increase the quantity of leaves and green vegetables fed, to increase the time that the rabbits spend eating, and then gradually reduce quantities of these. Grass is significantly more palatable than hay to most rabbits, so increasing the quantity of grass fed to the rabbits will also reduce the adverse effects on their welfare. The transition is often easier in outdoor rabbits kept on grass, as the grass cues normal grazing behaviour, and the rabbits are likely to just increase the time spent grazing, rather than having to substantially change the food they eat.

In conclusion, changing a rabbit's diet can cause stress and adverse effects on welfare, but the long-term gains, both behavioural and physical, outweigh these negative effects. Advice sheet 4 in the Appendix can help owners who are trying to improve their rabbit's diet.

Changing behaviour through behavioural modification

We've previously seen how owners can modify the rabbit's behaviour towards them by understanding the animal's motivations, responding to its behavioural signals and modifying their own behaviour accordingly. These are extremely important skills to improving the owner's relationship with their rabbit.

However, some unwanted behaviours may be very difficult, or even impossible, to resolve through changing the environment. In these situations, the owner can consciously modify the behaviour by training the rabbit. Training certain behaviours can help the rabbit to cope with the unnatural captive environment (as compared to the environment for which the wild rabbit is evolved). Additionally, training can be a form of environmental enrichment, and it strengthens the bond between the owner and the rabbit. Owners should be aware that the learning process is iterative and not linear, so they don't feel frustrated with their rabbit.

To understand how to train rabbits, owners need to understand how rabbits learn, which training techniques are most effective and how they can teach rabbits to perform desired behaviours.

Types of rabbit learning

Rabbits learn in two ways. These are called 'non-associative' and 'associative' learning.

Non-associative learning

Non-associative learning is a relatively permanent change in the strength of response to a single stimulus because of repeated exposure to that stimulus. It occurs when a rabbit is repeatedly exposed to a single stimulus, and that stimulus is not linked with anything else. There are two classes of non-associative learning.

- Habituation occurs when an animal becomes less sensitive to a stimulus to which it is repeatedly exposed.
- Sensitisation occurs when an animal becomes more sensitive to a stimulus to which it is repeatedly exposed.

Traditional recommendations for acclimatising rabbits to being picked up suggest that the owner should pick up the rabbit frequently. This relies on the rabbit desensitising to the unpleasant stimulus. However, if the interaction is sufficiently unpleasant, this 'desensitisation' process may well have the opposite effect: the rabbit may become more sensitised to it, and start to show escape or avoidance behaviours to prevent this interaction.

Non-associative learning can be useful in some situations, but where possible, any desensitisation process should involve some aspect of counter-conditioning: rewarding the rabbit for desired behaviour. Giving rewards can help to outweigh the unpleasant sensation of the stimulus to which the rabbit is exposed. Counter-conditioning (associating an unpleasant stimulus with a pleasant outcome) is an example of associative learning.

Some owners will think that their rabbit will 'get used' to being picked up if they pick it up frequently (habituation). Often, however, the rabbit becomes more anxious about being picked up (sensitisation), so will become harder to catch.

Associative learning

Associative learning, or 'conditioning', occurs when a rabbit learns that there is a link between an event and a behaviour. The rabbit learns either that it can predict an event (classical conditioning), or that it can control an event (operant conditioning).

- Classical conditioning occurs when an animal learns that a neutral stimulus (one to which the rabbit is indifferent) is paired with a potent stimulus (one that the rabbit is motivated by), and so it has an innate response when the neutral stimulus occurs. The well-known example is that of Pavlov's dogs: the sound of a bell was paired with food arriving, so the dogs would salivate at the sound of the bell. In classical conditioning, the response is involuntary or reflex.
- Operant conditioning occurs when an animal learns to modify its behaviour based

on the consequences of that behaviour, and learns to predict when to modify its behaviour based on what comes before it ('discriminative stimuli', or 'antecedents'). In operant conditioning, the rabbit changes its behaviour to achieve different consequences.

> *Rabbits salivate more when given food. If the owner always opens the fridge door to get vegetables for the rabbit, the sound of the fridge door predicts the arrival of the food. Over time, the rabbit will start to salivate when the fridge door is opened (classical conditioning). Subsequently, the rabbit may learn that if it waits by the fridge, the owner is more likely to open the fridge door and give it vegetables (operant conditioning).*

Classical conditioning

Classical conditioning gives rise to many responses that rabbits make to human behaviour. Diagram 4.4 shows how the sound of the fridge door is associated with the rabbit receiving food.

Classical conditioning doesn't often help when training behaviour, because the response is involuntary. However, it is used in clicker training to teach the animal that a 'click' is rewarding. Clicker training is a form of behaviour modification where the desired behaviour of the animal is marked by the sound of a click.

There is nothing inherently rewarding about the sound of a click: it is a 'neutral stimulus'. However, it is paired with a food reward, so becomes a 'conditioned stimulus' (like the sound of the fridge door in Diagram 4.4) – it predicts a reward. Once the rabbit has learned the association between the click and the food, subsequent behaviours are then trained using operant conditioning (see below). Clicker training will be discussed further later.

Operant conditioning

In operant conditioning, there are three elements that need to be learned:

- antecedent (stimulus)
- behaviour (response)
- consequence.

The nature of the consequence, in terms of the effect on the rabbit, predicts the likelihood of the rabbit showing the behaviour again in that situation. An animal only learns an association if the event affects the animal – either by causing fear or distress or by causing a positive experience. We call these motivators – an animal is motivated to perform a behaviour to avoid or achieve the experience (Diagram 4.5).

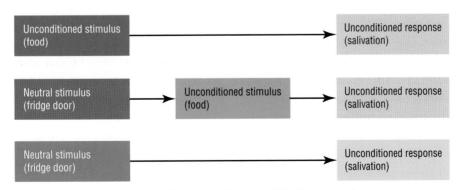

Diagram 4.4: Process of classical conditioning (rabbit starts to salivate when fridge door is opened).

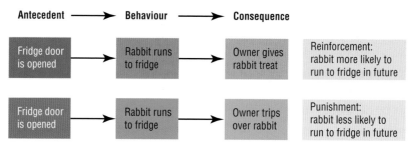

Diagram 4.5: Process of operant conditioning (rabbit runs to fridge when door is opened and receives food).

When the rabbit learns that it receives a reward (reinforcement) when it performs a behaviour, then it is more likely to perform that behaviour in future. If it learns that, if it performs a behaviour, there is an unpleasant consequence (punishment), then it is less likely to perform that behaviour in future. Consequences that increase the likelihood of a behaviour being repeated are called 'reinforcers', and those that decrease the likelihood of a behaviour being repeated are called 'punishers'.

Typically, reinforcement and punishment can be split into two categories each (see Diagram 4.6). If a reinforcer or punisher is given to the animal after it performs the behaviour, it is termed 'positive' (not because it is good, but because it is additive). If a reinforcer or punisher is taken away from the animal after it performs the behaviour, it is termed 'negative' (not because it is bad, but because it is subtractive).

Practically, for rabbit behaviour modification, understanding these different categories helps to understand why a rabbit has learned certain behaviours (Diagram 4.7). However, when undertaking deliberate training of a rabbit, positive reinforcement training is appropriate in the majority of situations. The one notable exception is constructional approach training, which uses negative reinforcement (the human backs away from the rabbit when the rabbit performs the correct behaviour).

The diagram gives an example of rabbit behaviours learned through positive and

	Something is given to rabbit after behaviour	Something is taken away from rabbit after behaviour
Behaviour more likely to be repeated	Positive reinforcement	Negative reinforcement
Behaviour less likely to be repeated	Positive punishment	Negative punishment

Diagram 4.6: Definition of reinforcement and punishment.

	Something is given to rabbit after behaviour	Something is taken away from rabbit after behaviour
Behaviour more likely to be repeated	**Positive reinforcement** When the owner calls the rabbit and it comes over, the owner gives it a treat.	**Negative reinforcement** When the owner reaches into the hutch, the rabbit bites the owner, and the owner stops reaching into the hutch.
Behaviour less likely to be repeated	**Positive punishment** When the owner calls the rabbit and it comes over, the owner picks it up.	**Negative punishment** When the owner calls the rabbit and it comes inside, the owner shuts the door so it can't go back outside.

Diagram 4.7: Examples of reinforcement and punishment.

negative reinforcement. These are all common situations, and in most of them the owner may not realise that they are training a behaviour.

Figure 4.15: Rabbits need to learn to take food from the hand before they can be clicker trained.

What motivates rabbits?

Different rabbits find different consequences reinforcing or punishing (for example, some may enjoy being stroked on the head, some may dislike it). In order to train a new behaviour, you need to understand what that specific rabbit wants to gain.

Common reinforcers are food, access to grass, access to a large space for exploration, and social interaction.

Common punishers are unpleasant sensations (being lifted, being turned upside down, being restrained, being frustrated), and being separated from a companion.

Owners usually feel that their presence or their interactions are reinforcers. However, in the case of many fearful or timid rabbits, their presence may be a punisher, and their withdrawal may be a reinforcer.

When a rabbit is relatively relaxed around humans, and will take food from the hand, then the owner can use food as a reward, and behaviours can be trained using a clicker. However, if a rabbit is very fearful, food will not be the major motivator – escape will be stronger. In this situation, constructional approach training (described later) is used to help the rabbit relax around humans. Once it is relaxed enough to accept food, the owner can move on to using clicker training (Figure 4.15).

Why is punishment usually ineffective in rabbits?

Case study 5: Pewter

Pewter was a three-year-old Polish male rabbit. He strongly disliked being picked up, and would bite his male owner when he lifted him. This behaviour became generalised, and he started to bite the male owner at other times.

The male owner did not appreciate this, and responded by punishing Pewter when he bit: he would catch the rabbit, restrain him and say 'No!'

Pewter changed his behaviour. No longer would he bite and wait to be caught. His new technique was to bite and then quickly run away.

The owner changed the behaviour by changing his interactions with the rabbit. He stopped picking Pewter up and used food rewards for good recall. Pewter gradually gained confidence with him, started to lick his hand when proffered (but did not like to be touched) and eventually began to seek interaction and enjoy having his head stroked. This took more than a month.

Humans intuitively understand punishment. When trying to stop an unwanted behaviour, we think of punishment before changing the motivation. There are arguments against using punishment on welfare grounds; however, the major reason to avoid its use when training rabbits is that it is rarely effective. There are seven reasons that punishment rarely works as intended with rabbits.

- Punishment is rarely specific enough.
- Punishment is rarely unpleasant enough.
- Punishment may increase the incidence of stress-related behaviours.
- Punishment may make the rabbit's behaviour unpredictable.
- Punishment slows learning.
- Owners are likely to escalate punishment.
- Punishment worsens the rabbit's relationship with the human.

Punishment is rarely specific enough

For a punishment to be effective, it must be situation independent (it should happen wherever the behaviour is performed), and consistent (it should happen every time the behaviour is performed), and should be specific (it only happens at the moment that the behaviour is performed). Therefore, there should also be no obvious source of punishment (i.e. if the punishment only occurs when the owner is present, the rabbit will learn that the owner is unpredictable, rather than that the behaviour is unwanted). In practice, it is very hard to achieve a punishment that is specific, consistent and independent of the situation. Many owners will use a loud noise or a word to deter a behaviour, such as 'No'. They often use the same punishment for multiple different behaviours (which is confusing), and the aversive effect of the word is very low (so the rabbit usually habituates to it). It may be enough to alert the rabbit, but not enough to stop the behaviour. Owners often report that the rabbit 'knows it is doing something wrong, but does it anyway'.

Punishment is rarely unpleasant enough

In order for a punishment to be effective, the aversive effect of the punishment must outweigh the rabbit's motivation to perform the unwanted behaviour. If the motivation is a strong evolutionary one (chewing, digging, finding food), then a punishment that was aversive enough to override them would have to cause substantial fear, distress or pain. This is inhumane.

Punishment may increase the incidence of stress-related behaviours, including aggression

If the behaviour is shown as a way to cope with stress, then punishment may increase expression of the behaviour. If a rabbit bites the owner, the owner may punish the rabbit by picking it up, by saying 'no' or by tapping it on the nose. A rabbit that bites from fear may bite more quickly, or bite harder, to try to avoid these unpleasant behaviours from the owner. Additionally, rabbits may start to show defensive aggression before the owner interacts with them to avoid a fear-inducing situation developing.

Punishment may make the rabbit's behaviour unpredictable

If a rabbit is punished when it lunges at an owner, or scratches them with its paws, then it may be less likely to show these behaviours again. However, if the fear- or frustration-inducing stimulus doesn't change, then the rabbit may escalate its behaviour and bite the owner without showing the warning signals of lunging or scratching. This makes the rabbit's behaviour harder to predict.

Punishment slows learning

Punishment is unpleasant by definition, which will cause mild stress. Stress causes the brain to increase activity geared towards fight or flight and decreases cognitive activity. This makes it harder for the rabbit to learn the desired behaviour pattern.

Owners are likely to escalate punishment

If punishment fails, owners are more likely to escalate the severity of the punishment than try another strategy (such as positive reinforcement). When owners punish a rabbit (i.e. do something unpleasant to the rabbit), they often feel frustrated or angry, which makes them more likely to worsen the punishment if it is not effective first time.

Punishment worsens the rabbit's relationship with the human

Any punishment that is associated with an owner increases the perceived unpredictability of the owner by the rabbit, which decreases trust. Where a species has an innate dependence on humans (such as dogs), punishment, although often not humane, may not cause such a breakdown in the relationship. However, in a prey species like a rabbit, with no inherent dependence on humans, when an owner uses punishment, the rabbit is much less likely to interact positively with the owner. Additionally, when punishment fails, the owner is likely to feel irritated with the rabbit, which will worsen their relationship with their pet.

How can positive reinforcement training be used with rabbits?

Positive reinforcement training can decrease stress, speed learning and increase the human–animal bond. The perfect reinforcement needs to be perfectly timed with the desired behaviour, and needs to be reinforcing enough that the rabbit wants to perform the behaviour to receive the reward.

The 'positive reinforcement' in this method of training usually involves food rewards for rabbits. As previously discussed, adult rabbits rarely play in the way that a dog does, so access to a toy is unlikely to be particularly rewarding. In addition, rabbits are not dependent on the owner to the same extent, so the interaction with the owner is rarely sufficient to reinforce a behaviour (the exception to this is when owners are allowing their rabbit to 'choose' whether it wants to be stroked. In this case, when the rabbit lowers its head, the owner strokes its face: the stroking reward is sufficient here because it is the reward that the rabbit 'expects' instinctively when performing this behaviour). While access to outdoors is motivating, it has the obvious disadvantage that once the rabbit is outside, the training session is effectively terminated (although this reward can be useful when an animal is learning to use a cat-flap or to move between different housing compartments). There are advantages to using food rewards: these can be small enough, and consumed fast enough, that the owner can give multiple rewards during one training session.

When a behaviour is being learned, the reinforcement should be consistent when the behaviour is performed correctly. Once the behaviour is learned, the frequency of reward for the behaviour should be reduced. If a rabbit has learned to recall to a command, the owner might only reinforce this behaviour every few times the rabbit performs it successfully. This helps to maintain motivation (see later).

What constitutes an appropriate food reward?

An appropriate food reward should be motivating enough that the rabbit wants to work to achieve it, but should not cause health problems when consumed in the quantities typical during a training session (especially when the sessions are daily). However, the most palatable foods (typically those high in sugar) are also the foods that should not be fed in large quantities. This means that, to protect the rabbit's health, the owner may need to change the diet so the appropriate food rewards become sufficiently rewarding. In practice, this usually requires the owner to stop feeding concentrate food in the cage, and instead to only give concentrate food during training.

Commercially prepared, high-quality concentrate, pelleted or extruded food pellets make great training rewards. This is because such diets are relatively balanced for rabbits, have grass as the major ingredient and the pieces are an appropriate size for a training reward (about the size of a pea). Concentrate food should constitute the major portion of training rewards (high-value 'jackpot' rewards are discussed later). An advantage of this approach

is that it encourages owners to view concentrate food as 'treats', rather than as a major dietary constituent, so may reduce the quantity that they feed their rabbit daily.

There may be challenges to using concentrate food for training rewards. If the rabbit has access to ad lib concentrate food (which, as discussed, is damaging to health), it will not be motivated by more concentrate food if this is offered as a reward. Owners may feel that they are depriving the rabbit of food solely to train it, rather than realising that this intervention is also important for the animal's health. Educating the owner on why and how they can transition their rabbit on to a more appropriate diet is therefore essential.

In order for the food rewards to be desirable (and appropriate), the rabbit's diet needs to be good. Concentrate food must not be fed ad lib, the rabbit should eat mostly grass or hay and any additional vegetables given should be green and leafy (so starchy root vegetables and sugary fruits should never be fed as part of the diet). There are two reasons for this: it is better for the animal's health, and restricting access to palatable, less healthy foods makes them more motivating. Grass and hay should never be restricted, as this increases the risk of gastrointestinal stasis. If an owner wants to train a rabbit using appropriate rewards, they first need to transition the rabbit on to a healthy diet.

If the rabbit is on a good diet, the quantity of pellets fed every day will be very small (if fed at all) – typically about a tablespoon per rabbit per day. Feeding more concentrate than this will reduce the rabbit's appetite for hay or grass (concentrate foods are designed to be very palatable – see earlier section on nutrition).

Owners should be counselled against using a lot of sugary rewards as treats during training. Rabbits like sweet tastes (an evolutionary advantage to detect higher calorie plants when grazing), and so are strongly motivated by sugary fruits (such as apple). However, they also have a higher concentration of taste buds than humans do, so will be motivated by the lower concentrations of sugar in root vegetables (i.e. carrot).

These fruit and vegetable treats are suitable for intermittent 'jackpot' rewards (see next section), but feeding them for all training sessions would result in overconsumption of these inappropriate foods. Owners may offer a wide variety of fruits to their rabbits in the mistaken belief that 'fruits are healthy', and this myth is perpetuated by the availability of 'rabbit food with tropical fruits', 'pineapple-flavoured treats' and other such marketed products. Small pieces (half the size of a fingernail) of sugary fruits or vegetables do have a role in training, but should constitute a small proportion of the training rewards (the majority being balanced concentrate foods).

> *The PAW report (2016) reported that a lot of owners gave treats to their rabbits to increase the variety of their diet.*

Increasing the value of food rewards

The less predictable a reward is, the more motivation the animal has to achieve it. This principle underlies the addictive nature of gambling. If a slot machine gave a penny every time it was played, the activity would be far less reinforcing than if it gave £20 intermittently. Predictable rewards allow a human (or animal) to assess whether or not an activity is worth the reward (and thereby reject the reward if the gain does not outweigh the cost), whereas unpredictable rewards prevent this decision-making process.

So, how can you make the reward schedule less predictable?

Make the frequency of reward intermittent

While the rabbit is learning a novel behaviour, the owner should reinforce the desired behaviour every time with food – this will ensure that the behaviour becomes consistent. However, once the behaviour is relatively consistent, the owner should start to reward every second or every third time, and then start to reward randomly. The rabbit can't afford not to give the desired behaviour because there is an unknown possibility of reward, and then when a reward comes it has more value because it does not happen every time.

Make the type of reward variable

Again, if the rabbit can predict what reward it will get, it can weigh up the costs of the behaviour it is being asked to perform against the predictable benefit it will get. However, if the reward is variable (mostly rabbit pellets, occasionally small pieces of carrot or apple), then it will be more motivated to learn.

Give occasional 'jackpot' rewards

As the owner starts to reduce the frequency of reward, the rabbit may decide that it is not worthwhile to perform the behaviour. However, if every so often you give a larger reward such as a piece of cabbage, the motivation to perform the behaviour will again be higher.

Training techniques

While altering the environment or husbandry is often sufficient to change a behaviour, training is very useful in some situations. Animals kept as pets usually live in environments that are quite different from the environment for which they are evolved. Therefore, training is most important when it helps the animal to cope better with its environment.

There are several reasons that training a rabbit can improve its welfare.

- Training a rabbit may decrease the frequency of unpleasant interactions with its owner. Training a rabbit to go into its cage on command, for example, prevents the rabbit from being herded or restrained to achieve the same goal, and will improve the owner–rabbit relationship.
- Training a rabbit provides mental stimulation. The majority of pet rabbit environments provide fewer opportunities to explore and learn than in the wild environment. This can be partly remedied by training 'tricks'.
- Training a rabbit improves the bond between it and its owner. The process of learning is reinforcing, and so the owner becomes a source of interest and reward for the rabbit. The process of successful teaching is also reinforcing for the trainer, so the rabbit becomes more rewarding to own. Additionally, reducing the negative interactions that the owner has with the rabbit will improve the relationship.

There are many different training methods. In rabbits, constructional approach training and clicker training are most appropriate.

Constructional approach training

Constructional approach training (CAT for short) is a form of training often used in horses and feral cats. It is a form of negative reinforcement training, where the reward is the withdrawal of the fearful stimulus (which in CAT is the presence of the owner).

CAT is useful for rabbits that are very fearful of their owners. When an animal is very stressed, the escape motivation overrides any motivation to find food, so such an animal can't be trained with food rewards. In this case, the biggest motivator is to get away from the human, and so this can be used as a reward.

The rabbits that respond well to CAT have typically learned the behaviour sequence depicted in Diagram 4.8.

When the owner approaches, the rabbit becomes more stressed. The owner does not perceive, or react to, the behavioural expression of this stress, and continues to approach. The rabbit is in 'fight-or-flight' mode, so either shows defensive agonistic behaviour (makes the owner withdraw), or runs away from the owner (the rabbit withdraws): both of these behaviours increase the distance between the owner and the rabbit, which reinforces the behaviour.

In CAT, the same reward is used (the rabbit is highly motivated to increase the distance between the owner and rabbit), but rather than

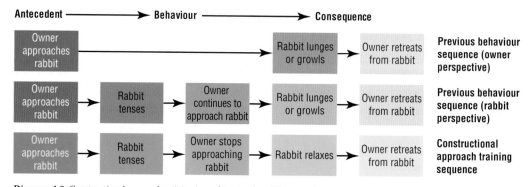

Diagram 4.8: Constructional approach training is used to give the rabbit control over its interactions with the owner by showing calm and relaxed behaviours rather than fearful or aggressive ones.

the rabbit achieving the reward through either a 'fight' response or a 'flight' response, the owner withdraws when the rabbit remains calm.

In CAT, it is really important to help owners learn the early signals of stress in their rabbits. Once they can identify these, they can start to adjust their own behaviour based on the behaviour of the rabbit. The trigger (the approach of the owner) must always below the threshold for the fight or flight response.

This technique has been demonstrated to be effective in a variety of species, including birds and even reptiles. Interestingly, reinforcing calm behaviour changes the emotional state of the animal to the trigger: the perception of control builds confidence. Typically, animals trained using this method start to come forward to investigate and interact with the previously aversive trigger.

CAT consists of the following steps:

- setting a baseline for approach
- approach
- switchover
- interaction
- generalisation.

Let's explore these further.

Setting a baseline for approach
When starting CAT, the owner needs to find the distance from their rabbit at which the rabbit can cope with their presence without running away or attacking (the behaviours that are unwanted).

Approach
The owner moves their hand to the baseline level (rabbit is likely to tense body or face, or flatten ears), and waits for the rabbit to show a calm behaviour (shift in body position, ears up, movement). As soon as the rabbit does this, the owner moves their hand away. The owner

waits 10–15 seconds before moving their hand to the baseline level again. The rabbit should show a calm behaviour more quickly (desensitising to the approach, learning the reinforcement pattern). The owner should withdraw their hand as soon as the rabbit shows the calm behaviour. After a number of repetitions, the owner can start to gradually increase the amount they advance their hand, ensuring that they withdraw as soon as the behaviour is seen.

Switchover
At this point, the rabbit should be showing the desirable calm behaviours more frequently and more quickly, and the owner should be able to get quite close to the rabbit. During this period, the behaviour has been reinforced and is therefore expressed more quickly, but the emotional state is still fearful. It takes longer to change the emotional state than the behavioural expression.

Interaction
Once the emotional state of the rabbit starts to change, from fear to curiosity, the rabbit will start to try to interact with the trigger. This is a signal that the constructional approach training has been successful (although if the rabbit shows more anxiety behaviours, the owner needs to move back a stage). When the rabbit shows that it wants to interact, its motivations have changed: distance is no longer reinforcing, but interaction has become reinforcing. This means that if the owner withdraws when the rabbit wants to interact, they are removing the reinforcer – an unintended punishment! When the rabbit wants to interact with the hand, initially the owner should stay still and not try to touch the rabbit. Once the rabbit becomes more confident, the owner should offer food rewards from the hand, positively reinforcing the rabbit's approach.

Generalisation

Initially, when a new behaviour is learned, it is very specific to the environment in which it has been learned. The final stage of training a behaviour is ensuring the animal can generalise this behaviour to different situations. Owners should try approaching the rabbit with their hand when it is in different places, and following the above steps. This process should be substantially quicker once the rabbit has learned it in one place. If the rabbit approaches the owner in a calm way in different situations, then the training has been successful.

Advice sheet 7 in the Appendix can help owners with constructional approach training.

Clicker training

Clicker training is a term for a type of predominantly positive reinforcement training that uses a 'bridging stimulus' to mark the desired behaviour. A bridging stimulus is a neutral stimulus (like a click, made by a device called a clicker) that predicts a reinforcer (a food reward). The click becomes a 'conditioned reinforcer' – although it has no intrinsic value, it is associated with something rewarding. During the association of the clicker with the food reward, the rabbit learns that a 'click' means the behaviour is correct (Diagram 4.9).

Clicker training is not essential to train any behaviour: simple behaviours can be trained using food rewards without a bridging stimulus. The value of clicker training is that the conditioned reinforcer is very specific in terms of time (it is easier to give a click at exactly the right time than it is to immediately sprint to the rabbit to give it a food reward), it is very specific in terms of sound (the rabbit won't hear that sound in any other context) and it can be used to mark good behaviours in a behaviour chain without interrupting the sequence to give a food reward (Figure 4.16).

The larger clickers (designed for dogs) make a loud noise, which many rabbits find unpleasant: if the rabbit is initially fearful of the noise, their learning will be much slower. Small clickers can be purchased; alternatively, the pop-up lid on a jam jar can also be used.

Clicker training consists of the following steps:

- association of click with treat
- association of desired behaviour with treat
- putting the desired behaviour on cue.

Let's explore these in more detail.

Association of click with treat

This first step is to teach the rabbit that 'click means treat'. In order to learn this, the rabbit

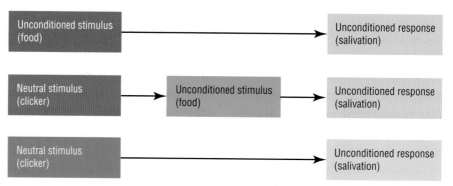

Diagram 4.9: Clicker training uses a combination of classical and operant conditioning. The first step is to teach the rabbit that the click predicts food.

Figure 4.16: Many dog clickers are not suitable for training rabbits, as the sound is too loud. Some clickers, such as Karen Pryor's iClick clicker, produce a sound that is quieter.

must be confident in taking food from the hand. If this is not the case, the owner should start to hand feed the rabbit's concentrate food ration. The owner should do this daily for at least a week so the rabbit is not fearful or stressed in this situation. Initially, it may be easier to feed the rabbit with a portion of concentrate food in the cupped hand, but the owner should then move to feeding the rabbit with individual pieces of food in the fingers (when using food rewards for training, the goal is to give enough food to provide a taste reward but not enough to make the rabbit feel full after performing one behaviour!).

When the rabbit is easily hand fed, the owner can then start to teach an association between the click and the treat (sometimes referred to as 'charging' the clicker, or using classical conditioning). The owner should give a click and then immediately give a food reward, ensuring that they do not make any movement to give the reward until the sound is made (otherwise the rabbit may associate the movement, rather than the click, with the reward). The owner should repeat this sequence, click then treat, click then treat, until the rabbit starts to visibly react when it hears the click: usually by looking at the hand or moving slightly forward.

At this point, the rabbit has been classically conditioned: the sound of the click predicts the food reward.

Association of desired behaviour with click

The second stage is to teach the rabbit that it can 'cause' a click by performing a desired behaviour. This association can either be taught by clicking and rewarding a behaviour that the rabbit already knows is a 'good behaviour', or by clicking and rewarding a behaviour that is simple and performed frequently.

An easy behaviour to click and train is approach. A rabbit that knows that its owner may give it a food reward is likely to approach the owner. The owner can click and reward approach behaviours, and can move away from the animal to encourage the rabbit to show this behaviour.

Reinforcing an initial behaviour is easy, but the association may take a while for the rabbit to learn. Once the rabbit has experienced a 'eureka moment', i.e. it has learned the association between the behaviour and the report, then the behaviour will become more frequent and more specific. Once the rabbit has learned one behaviour through clicker training, subsequent behaviours can be learned much more quickly.

Behaviours can be taught in several different ways.

1. Luring
    'Luring' is commonly used to teach a new behaviour. The owner holds a food reward in front of the rabbit, and moves it until the rabbit is in the right position (Figure 4.17).

When the rabbit performs the correct behaviour, the owner clicks and rewards it.

An owner might lure the behaviour of the rabbit standing on its hind limbs by holding food rewards above the rabbit's head. When the rabbit stands on its hind limbs, the owner clicks and rewards it. The lure is then faded out when the behaviour is put on cue (see next section).

2. Capturing
    'Capturing' a behaviour refers to clicking and rewarding a behaviour that is offered spontaneously. An owner might capture the behaviour of the rabbit standing on its hind limbs by clicking and rewarding the rabbit when it stands on its hind limbs while exploring its environment. Capturing can be a useful way of training a behaviour, but can only be used on behaviours that are offered spontaneously by the rabbit.

Figure 4.17: New behaviours can be trained using 'luring'. In this case, the rabbit follows the food reward, puts his paws on the box, and then the reward is released.

Figure 4.18: This rabbit has been trained to touch his nose to the blue target. Once this behaviour is trained, new behaviours can be taught – the rabbit will follow the target.

3. Shaping
    'Shaping' a behaviour refers to rewarding incremental approaches to the desired behaviour.

    An owner might 'shape' the behaviour of the rabbit standing on its hind limbs by clicking and rewarding the rabbit every time it raises its head. As the rabbit becomes more confident in offering this behaviour, the owner will wait to click so the rabbit raises its head more each time. Eventually, the rabbit may lift its forepaws from the floor and the desired behaviour will have been achieved.

    Shaping can be very useful to train behaviours that are not offered spontaneously, but the behaviour may take longer to train.

4. Targeting
    'Targeting' a behaviour refers to teaching the rabbit to touch a 'target' with its nose as an initial behaviour (Figure 4.18), and then teach subsequent behaviours by moving the target so the animal performs the correct movement or position.

    An owner might use targeting to train a rabbit to stand on its hind limbs by teaching it initially to touch its nose on to a target, and then by holding the target over the rabbit's head.

    There are some similarities between training an animal using a lure or using a target: in both of these, the animal is following an object that may directly or indirectly confer a reward. It can be difficult to phase out the lure or the target when the behaviour is trained. The motivation of the animal may be different: some trainers find that it is slower to train animals through use of a food lure than use of a target, as in the former, the animal is focused on the reward rather than focused on the task.

Putting the desired behaviour on cue

It may be useful to increase the frequency of a desired behaviour, but the real value comes when the rabbit can perform a behaviour 'on cue', i.e. when the owner gives a signal. An owner may like it if the rabbit approaches them more frequently, but the owner will find it more useful if the rabbit approaches them when they call it.

Certain types of cue are easier for rabbits to learn than are others. Rabbits primarily communicate visually, so they tend to find hand signals easier to learn than vocal commands. Rabbits have not evolved for high visual resolution and spatial perception at short distances (the eyes are large and the visual fields do not overlap). Visual cues need to be large and specific enough for the rabbit to perceive them and differentiate them from other similar gestures. Vocal or auditory commands should not cause fear (so not overly loud or intimidating) and should be specific. Tactile and olfactory cues can also be learned, but are rarely useful for trained behaviours.

When a rabbit is predictably performing a behaviour to obtain a click and reward, the owner can then start to train a cue. At that point, the rabbit has probably already learned the owner's unconscious cues (when the owner kneels on the floor, they are likely to reward the rabbit for coming towards them). The owner should give the cue when the rabbit looks like it is about to perform the correct behaviour, and then click and reward the rabbit when it performs that behaviour. If the owner is moving towards a vocal cue, they should use the vocal cue before they offer the hand signal, to train the rabbit that the vocal cue predicts the visual cue for the behaviour.

What cued behaviours are useful in rabbits?

The major reason to train a rabbit is to teach behaviours that allow it to have a better quality of life. There are some cued behaviours that are essential (for managing the rabbit, reducing harm if the rabbit escapes and interacting with the rabbit in a mutually rewarding way), and others that are optional (may not be necessary because of environmental set-up, or may be difficult to train and therefore not suitable for an owner who lacks motivation). Clicker training can be used to teach some of these behaviours; for some behaviours there are other motivators that can be used.

Essential behaviours include:

- recall
- acceptance of interaction
- entering transport container
- standing on hind limbs.

Let's investigate these further.

Recall

The rabbit comes to the owner when the owner gives a cue. This behaviour is essential: it allows the owner to interact with the rabbit without chasing it, it allows the owner to move the rabbit between places without catching it and it is a vital behaviour to limit risk if the rabbit escapes. The behaviour is easily trained with food rewards: the owner should use a predictable cue (such as a whistle) before offering food. Once the rabbit reliably responds to the cue by coming over to the owner, the owner should gradually increase the starting distance between them. This behaviour will be harder to train if the owner frequently picks up the rabbit, as it will be less inclined to approach the owner.

Acceptance of interaction

The owner gives a cue when he or she would like to stroke the rabbit, the rabbit chooses whether to interact and the reward is provided when the owner strokes the rabbit. This increases the predictability of the owner–rabbit interactions and builds trust; it also helps the owner to focus on observing the rabbit's behaviour and responses and adjusting his or her own behaviour appropriately. This behaviour is also easy to 'train' as it is a common social behaviour between rabbits, and the 'reward' of facial grooming is expected by the rabbit.

Owners should train this behaviour by waiting until the rabbit is calm and relaxed (especially if it has a history of aggression) and putting their closed fist on the floor in front of the rabbit's face. If the rabbit remains relaxed, they can gently push their fist forwards until it is in contact with the rabbit's whiskers. The owner should leave it there for a few seconds: if the rabbit wishes to be stroked, it will lower its head; if it doesn't change head position, the owner should back off and try again a few minutes later. If the rabbit wishes to be stroked, the owner should stroke and rub between the rabbit's eyes and over its face.

Owners can use this cued behaviour as part of a health check. As the rabbit becomes more confident, the owner can rub along the eyelids to remove tear crusting and palpate along the cheek teeth in the upper and lower jaws.

This behaviour should not be reinforced using food rewards, because the goal is that the rabbit enjoys a non-food-based interaction. Giving treats for a behaviour that is intrinsically pleasurable decreases an animal's motivation to perform that behaviour.

Entering transport container

The rabbit enters the transport container when the owner gives the cue. This behaviour is important to allow the owner to transport the rabbit in the car or to the veterinary surgeon. It can be a difficult behaviour to train, as the motivations of the food reward must outweigh the rabbit's instinctive dislike of restraint. The learning process should be as follows: habituation to the container (container left on the floor for the rabbit to investigate), food left in container for rabbit to find (not associated with owner: important if rabbit associates previous aversive experiences with the container) and then positive reinforcement by the owner with food rewards when the rabbit enters the box.

Once the rabbit is consistently going in and out of the box on cue, the owner can then touch the door before giving a food reward, move the door and give reward, close the door and give reward, and then move the box containing the rabbit over a short distance. As the feeling of restraint is unpleasant to the rabbit, this process should be very slow, and the owner should wait until the rabbit is comfortable at every stage before proceeding.

Standing on hind limbs

The rabbit stands on its hind limbs when the owner gives the cue. This behaviour does not necessarily need to be trained, as a rabbit that is used to taking food from the hand will willingly stand on its back legs to reach the food. This behaviour is important, however, as it allows the owner to assess the abdomen and anogenital area of the rabbit without needing to invert it (checking for matting, faecal soiling or injury).

Optional behaviours include:

* training a rabbit to be lifted
* paw contact
* tricks.

Training the rabbit to be lifted

Even with the best environmental set-up, there may be a few times in the rabbit's life where it will need to be picked up. It is possible to train a rabbit to accept this, but the rabbit must trust the owner and the reward must outweigh the unpleasant experience.

To give the rabbit a choice over whether or not it is being picked up (essential if you're doing something unpleasant to it), the owner should teach a 'request' cue. This signals to the animal what your intention is, and the animal can then choose to comply or to retreat. The aim is to reduce the unpredictability of the owner's interactions with the rabbits. There are several methods that can be taught, depending on the rabbit's previous experience and dislike of being lifted.

There are various components of being lifted that the rabbit may find unpleasant: being restrained, unaccustomed human contact on sensitive body areas and lack of paw support. During this training process, the owner should accustom the animal, as much as possible, to each unpleasant sensation in sequence: they should gradually increase the intensity of the stimulus.

Lifting a small rabbit using the hands

Rabbits often strongly dislike the sensation of human hands on their paws. Small rabbits can be picked up by 'cupping' them (applying pressure along the sides of the body with the hands slightly curving underneath) (Figures 4.19 a, b, c). This negates the need to actively reach under the rabbit, so is perceived as less unpleasant and is therefore easier to train. It is only suitable for short-bodied rabbits (depending on the size of the owner's hands, longer-bodied rabbits cannot be supported adequately with this method).

Initially, the owner should use a hand signal in front of the rabbit. This cue must be large enough to be perceived (the eyes on a rabbit do not face forward), and must be specific only to the intention to pick the rabbit up. One suitable gesture involves the hands opening like a book and moving on to each side of the rabbit's face.

A rabbit that trusts its owner is likely to lower its head. This behaviour, in response to this cue, is going to become the 'acquiescence' signal. When the rabbit lowers its head, the owner should bring the hands gently on to each side of the rabbit, briefly touch its flanks, then click and

Figure 4.19 (a, b, c): Initially train the rabbit to accept hands on each side of its body. Then gradually increase the pressure, then finally move on to lifting the rabbit off the floor.

Figure 4.19 (a, b, c): Continued

reward and move the hands away. This interaction will be unexpected; the rabbit is likely to become more alert at the unusual interaction and the unexpected reward. At the next trial, the rabbit may lower its head or may keep its head up: if the latter, move the hands away; if the former, repeat the behaviour and the reward.

As the rabbit becomes more confident in the behaviour, the intensity can be increased: mild pressure can be applied along the body, and if the rabbit is still acquiescing to the interaction, then the weight can be incrementally lifted off the paws, until the rabbit can be lifted and moved.

This training does not teach a rabbit to enjoy being lifted, but allowing the rabbit control over this sensation helps it to tolerate the restraint in exchange for the reward. As the sensation is unpleasant, owners should be aware that they will not be able to reduce the frequency of reward much without the behaviour itself becoming less reliable.

Lifting a large rabbit using the hands

Larger rabbits cannot be lifted using the above method. The alternative method for training the rabbit to be lifted using positive reinforcement has similar principles to the above method, but involves a higher level of unpleasant stimulation and so is harder to train.

The 'request' cue should be trained as above, but rather than moving the hands along the sides of the rabbit, one hand should be moved to touch one of the shoulders of the rabbit. As the rabbit becomes more confident, the hand can apply more pressure, and then gradually move under the rabbit to support it under the sternum. Once the rabbit is confident with this (continuing to acquiesce to the interaction without retreating), the other hand can be used to apply mild pressure on the rump. Ideally, this hand would support the hind paws, but this is very rarely tolerated: it may be easier

to cup the rump of the rabbit above the tail, lift the forelimbs and use forward pressure on the rump to lift the rabbit. The rabbit should not be held in this position for a long period, but this method can be used to transfer a rabbit into a box, for example.

Some rabbits will not tolerate this at all. These rabbits are better candidates for being lifted in a towel.

Lifting a rabbit using a towel

This method reduces the aversive stimulation of being picked up by avoiding any contact between human hands and the rabbit's paws. While rabbits may find insecurity under the paws unpleasant, this seems to be perceived as less unpleasant than contact with human hands. This method is often easier to train than either of the above.

As above, the intensity of the stimulus is gradually increased as the rabbit becomes more confident. In this situation, the 'acquiescence' behaviour is the rabbit hopping on to the towel to be lifted: if it does not hop on to the towel, it will not be lifted.

Initially, the owner should place a small cushion on the floor and cover it with a medium-sized towel. The cushion stays fairly flat when picked up, so will provide more stability for the paws while the behaviour is being trained. The end goal of the training is that the rabbit can be lifted in the towel without the cushion. The owner should click and treat the rabbit when it hops on to the cushion, and then should start to increase the time between the rabbit getting on to the cushion and the click, to encourage the rabbit to wait in this position. During this 'wait' time, the owner should start to gently lift the two edges of the towel – initially by a few centimetres, and then further.

The next increment is to put the hands under the towel and start to shift the cushion slightly, to accustom the rabbit to insecurity of footing.

Figure 4.20: Training a rabbit to put its paws on the owner's hands allows the owner to assess claw length.

As the rabbit becomes confident and stays still on the cushion while it moves, the cushion can start to be lifted. Then the edges of the towel can be lifted to block the rabbit's view while the cushion is moved. Then the edges of the towel can be lifted to take some of the rabbit's weight. Finally, the towel can be lifted without the cushion. It is very important to respect the rabbit's choice and reward desired behaviour at every stage.

If an owner struggles to train any of these behaviours, they should be advised to find other ways to manage the rabbit that it finds less unpleasant. Training it to go into a box on command, or setting up the environment to avoid the need to lift the rabbit, are both preferable to lifting the rabbit.

Paw contact

Training a rabbit to give a paw or put its paws on the owner's hands is useful as it allows the owner to assess claw length without lifting the rabbit. This behaviour is easy to train once the rabbit is used to being hand fed: the owner should place their hand in front of the rabbit and use a piece of food above the rabbit's head and above the owner's hand. The rabbit should rear on to its back legs, and lean forward to get the treat: to balance itself, it is likely to put its paws on the owner's hand. The owner should then click and treat this behaviour. A variation is to offer food at the top of the hand and well above the rabbit's head: the rabbit will usually 'grab' the hand between its paws to reach the food (Figure 4.20).

Tricks

Training tricks may seem pointless, from a health and welfare standpoint. However, there are various secondary benefits that can be very powerful. For example, tricks are rewarding for owners, which will make training a rewarding experience and increase the likelihood of owners choosing to use it for other purposes (like those above). Also, owners are often

Figure 4.21: Rabbits can be trained to scratch their claws on sandpaper – this reduces the need to handle them to clip the claws (which most rabbits dislike).

proud of their rabbits when they learn tricks, which means both a stronger human–animal bond and the possibility of the owner showing other (current or future) owners that rabbits are capable of learning these behaviours if kept correctly. Society learns through these sorts of interactions.

Rabbits can learn a variety of different tricks. These may be cued by the owner (using a hand signal or a voice signal) or may be cued by the presence of an object. Examples of owner-cued 'tricks' are: going out of a room on command, going into the cage on command, walking on the hind legs, following a point (owner indicates in what direction the rabbit should move to find the reward), a sequence of agility behaviours or nodding the head. Examples of object-cued behaviours are: scratching a block covered in sandpaper (to wear down the claws, Figure 4.21), throwing an object, placing an object in another object, retrieving an object or performing a behaviour that causes a noise to be produced. There are various methods used

to train these behaviours, which are covered in the next section.

Tips for improving clicker training efficiency

While clicker training is simple and effective, owners may encounter a variety of problems as they start to train their rabbit. Anticipation of the likely causes of these problems ensures that the principles of training are well understood and communicated to the owner, and that any problems encountered can be addressed and resolved.

'The rabbit doesn't seem to be learning'

If the owner doesn't feel that the rabbit is learning, there are four aspects of the training that should be considered: timing, criteria, rate of reinforcement and quality of reinforcement.

◆ **Timing:** Check that the owner is timing the click to coincide with the desired behaviour. If they are early or too late, the rabbit will associate another behaviour with the reward.

- **Criteria:** Find out what the owner considers 'success'. When training a behaviour, the owner should initially reward approximations toward the correct behaviour, and then gradually select for a better and better behaviour. If their criteria are too strict, so they only reward a perfect behaviour, then the rabbit is likely to get bored. If their criteria are too lax, and they continue to reward low-level approximations of the behaviour as the rabbit progresses, the rabbit won't learn the correct behaviour. As a rule, the owner should set achievable criteria ('I will click and treat the rabbit when it comes over to me from a metre away'), and then should not change the criteria until the rabbit is performing this behaviour for at least 80 per cent of the repetitions. Once the rabbit meets this threshold, the owner can tighten the criteria ('I will click and treat the rabbit when it comes over to me from two metres away').

- **Rate of reinforcement:** While the rabbit is 'learning how to learn', the training sessions should be very rewarding. The rabbit should be earning a click every three to five seconds when trying to perform a behaviour on cue: if the rewards are less frequent, the rabbit is likely to get bored. Additionally, the training sessions should be rewarding but short: any longer than five minutes and the rabbit is unlikely to be able to concentrate. The owner should always finish a session with a successful behaviour (so they may need to use a cue that the rabbit knows well).

- **Quality of reinforcement:** When training a new or difficult behaviour, the quality of the treats needs to be higher to provide the motivation to learn the behaviour. Good 'standard' treats are small pieces of concentrate food: these are small enough to provide a taste reward without a satiety reward. However, the strength of the reinforcer can also be altered by intermittent 'jackpot' rewards: these are very high-value rewards used occasionally when the rabbit performs a behaviour very well or very quickly. Examples of 'jackpot' rewards include a piece of apple peel or carrot (if the rabbit likes these foods), or even several pieces of concentrate food (so a quantity, rather than variety, reward). When a rabbit unpredictably receives 'jackpot' rewards, it will be more motivated to perform cued behaviours as it is unable to weigh up the difficulties of performing the behaviour with the reward of the treat, as it does not know what that treat is going to be.

'The rabbit isn't interested'

Simplistically, rabbits will be most interested in the most interesting aspect of their environment. Therefore, to increase the rabbit's ability to concentrate on a training session, either the training session needs to be made more interesting than the environment, or the environment needs to be made less interesting than the training session.

To increase the rabbit's interest in the training session, owners can increase the rate at which the rabbit receives treats, the relative value of the treats or the variety of the treats.

To decrease the rabbit's interest in the environment, the owner should ensure that the environment in which they first start to train a behaviour is free from distractions (recall over short distances in a room). Once the rabbit performs the behaviour reliably, then training can be extended to different situations with more distractions (recall across a garden).

'I can't see how to achieve the behaviour change I want through clicker training'

Not all behaviours are amenable to being changed through clicker training techniques. Some behaviours are normal behaviours, and the motivation to perform them is so strong that the owner cannot provide a stronger motivation to perform another behaviour (for example, if the owner tried to train a rabbit not to eat its caecotrophs).

Assuming that the behaviour can be trained, the owner should think about how they can break down the behaviour into small steps that can be reinforced, and how they can set up the rabbit to perform each step of the process (see section on Clicker training at p. 143).

'How do I train a behaviour sequence with several different steps?'

Slightly counter-intuitively, if the owner is trying to train a behaviour chain, they should first train the last behaviour in the chain. Imagine that the owner wants the rabbit to perform a sequence of behaviours along an agility course (slalom, tunnel, jump, ramp). They should train the rabbit to go over the ramp first. Once the rabbit is confident going over the ramp, then the owner can train the rabbit to go over the jump before tackling the ramp. And then extend the sequence again once the rabbit is confident.

This means that the rabbit is always progressing towards a behaviour that it is more confident in performing. For this reason, to train a rabbit to recall, it is easier to train the rabbit to come to the hand from a short distance away, and then gradually increase the distance. If instead, the rabbit was reinforced for taking a step towards the owner when it was a long way away, it would be much more confident about taking the first step in the correct direction than actually coming to the hand.

'How do I stop using the clicker?'

The clicker is useful for training the behaviour, as it marks the correct behaviour very specifically. However, once the behaviour is learned, the clicker no longer has a function, so can be phased out. Initially, the rabbit can be given a reward when the behaviour is performed but without a preceding click. Then the owner should make the reward schedule intermittent, i.e. reward the behaviour every other time or every few times the animal performs it. However, they should also make the rewards bigger: so an intermittent reward for a good recall might be a small handful of food scattered on the ground. If an owner completely stops reinforcing a behaviour, then the rabbit will become less likely to perform the behaviour. However, intermittent reinforcement will increase the rabbit's motivation to perform the behaviour. Owners can learn to mimic the noise of the clicker with their mouth for intermittent rewards, which will free them from carrying the clicker.

Changing behaviour through pharmacological means

There are some pharmacological products licensed in dogs and cats for specific types of behaviour modification, but these are only licensed in conjunction with a behaviour modification plan. None of these drugs are licensed in rabbits for behavioural problems, and there is no evidence to suggest that they are effective. Pharmacological methods of changing behaviour in rabbits should not be recommended.

5

How can I resolve a specific behavioural problem?

So far, this text has covered how to diagnose the underlying causes of a behavioural problem and how to train a new behaviour. This section explores how to prevent a problem behaviour. Some of these behaviours are normal behaviours that cause a problem to the owner; some of these behaviours are abnormal behaviours that cause a problem to the owner. This chapter will cover how to manage unwanted behaviours: initially by using general principles, and then by addressing specific behavioural problems.

Once you've worked out why an unwanted behaviour is happening, and you've identified the behaviour that you want to happen instead, then you and the owner need to work out the best way of getting the rabbit to perform the desired behaviour. Throughout this process, it is very important to get the owner involved in the diagnostic process and then the treatment plan. If they understand what the rabbit is trying to achieve by the behaviour that it is performing, they are more likely to feel empathy than frustration, which will make it easier for them to change their own behaviours to get quicker resolution.

The least effective behavioural modification plan is the one that is not acted upon. An actionable plan should be simple for the owner to implement and should fit into the owner's routine. The easiest way to achieve this is to involve the owner in the construction of a suitable behavioural modification plan. Explain the problem, describe the 'ideal' solution, and then ask the owner how they might try to achieve a similar result.

It is easy to assume what the owner sees as barriers to resolving this behaviour, but you may be surprised. Owners may also feel that, when they understand a behaviour, they can 'fix' it in another way. Their approach may be suitable or unsuitable, but discussing it openly allows you to explain why their approach may or may not work.

> 'Your rabbit is defecating on the floor because he is marking his territory. This is frustrating, especially because it is soiling your carpet. Many owners manage this by restricting the rabbit's access to a single room or space overnight. What do you think you could do to manage this behaviour?'

When an animal is motivated to repeat a behaviour many times, the behaviour becomes harder to change. This is because the strategy becomes a habitual response (so becomes more automatic), the degree of response may escalate (so becomes more severe) and the behaviour may start to occur in more situations (so becomes generalised). For example, it can be hard to reduce a rabbit's aggressive response to being approached when it is in its hutch, but it is much harder if the rabbit shows aggression whenever and wherever it is approached.

Resolving an unwanted behaviour

There are various ways to prevent any unwanted behaviour. The more times that an animal performs the behaviour in a given situation, the more habitual it becomes, and the harder it is to modify or stop the behaviour.

Karen Pryor brought the idea of clicker training to the general public through publication of her book, *Don't Shoot the Dog!* In it, she outlines the eight ways of preventing an unwanted behaviour from occurring.

- Method 1: Remove opportunity for the rabbit to show the behaviour.
- Method 2: Punish the behaviour.
- Method 3: Negatively reinforce the behaviour.
- Method 4: Wait for the behaviour to go away on its own.
- Method 5: Train an incompatible behaviour.
- Method 6: Put the behaviour on cue.
- Method 7: Shape the absence of the behaviour.
- Method 8: Change the motivation of the rabbit.

Different behaviours require different solutions, but as a rough guide, the eight techniques are ordered from least effective and least humane to most effective and most humane. The first four are negative methods, and the second four, positive methods. When preventing or modifying an unwanted behaviour, all solutions will involve one or a variety of these methods. We'll now consider each of these methods in turn, and when they may be appropriate to use when training a rabbit.

Method 1: Remove opportunity for the rabbit to show the behaviour

Method 1 refers to any method that physically prevents the rabbit from performing the

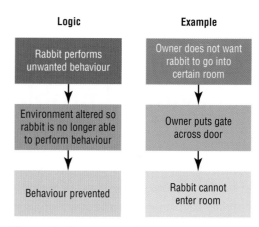

Diagram 5.1: Remove opportunity.

behaviour (see Diagram 5.1). In extreme cases, the owner could get rid of the rabbit: this will stop the unwanted behaviour from occurring. Other examples might include: the owner who can't stop their house rabbit chewing the furniture, so decides to keep the rabbit outside in a hutch; the owner who struggles to litter train their rabbit, so restricts them to a small area; or the veterinary surgeon who advises long-term use of an Elizabethan collar to prevent a rabbit mutilating itself.

In each of these examples, the rabbit does not learn from the behavioural intervention. Usually, this results in some form of welfare compromise: restricted access to space, or inability to perform a coping behaviour.

There is a place for Method 1 during rabbit training, as a short-term intervention to prevent a behaviour while alternative behaviour patterns or alternative husbandry systems are being put in place, but in general it should be avoided.

Method 2: Punish the behaviour

Diagram 5.2 shows the default behavioural modification process that humans think of. However, as discussed in the previous section, punishment is extremely hard to get right, usually has other consequences and is rarely

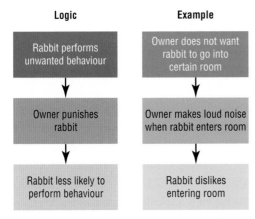

Logic	Example
Rabbit performs unwanted behaviour	Owner does not want rabbit to go into certain room
↓	↓
Owner punishes rabbit	Owner makes loud noise when rabbit enters room
↓	↓
Rabbit less likely to perform behaviour	Rabbit dislikes entering room

Diagram 5.2: Punish behaviour.

effective at changing a behaviour. Additionally, when an owner uses punishment unsuccessfully, they are more likely to try using a more severe punishment than they are to try switching to another method.

While punishment may stop an unwanted behaviour in the short term, it does very little to prevent the behaviour in the long term. The rabbit doesn't learn what it should do, but it does become more stressed, and usually more fearful of the owners (which can even lead to worsening of stress-related coping behaviours). Additionally, owners who punish their animals are more likely to feel unhappy, both because of the guilt of doing something unpleasant to an animal, and because of frustration with the lack of response.

Rabbits seem to learn very poorly when trained by punishment: probably because the fear and stress created impede the ability to form memories. The fearful rabbit's drive is to escape from the unpleasant situation, but their ability to associate the fear with their prior behaviour seems fairly limited. Punishment is very rarely useful when managing unwanted behaviours in rabbits.

Method 3: Negatively reinforce the behaviour

Horses are usually trained using negative reinforcement – applying an unpleasant stimulus

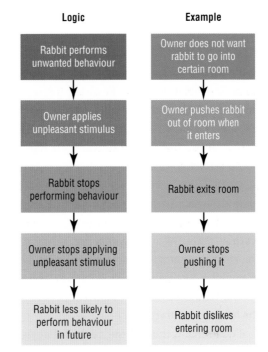

Logic	Example
Rabbit performs unwanted behaviour	Owner does not want rabbit to go into certain room
↓	↓
Owner applies unpleasant stimulus	Owner pushes rabbit out of room when it enters
↓	↓
Rabbit stops performing behaviour	Rabbit exits room
↓	↓
Owner stops applying unpleasant stimulus	Owner stops pushing it
↓	↓
Rabbit less likely to perform behaviour in future	Rabbit dislikes entering room

Diagram 5.3: Negatively reinforce behaviour.

until the desired behaviour is performed, at which point the unpleasant stimulus is stopped to reinforce the desired behaviour. When a rider wants a horse to turn to the left, they apply unpleasant pressure on the left-hand side of the horse's mouth through the bit. The horse turns its head to the left to prevent the unpleasant pressure, and the rider releases the pressure. In the future, the horse will turn its head faster to avoid the pressure: this behaviour has been trained through negative reinforcement.

If an owner wants a rabbit to move away from an area, he or she might try to push it (Diagram 5.3) or move it with their feet (especially if they are concerned about being bitten). This is a form of negative reinforcement – the rabbit finds the sensation of being pushed unpleasant, and when it moves away, the unpleasant sensation stops.

However, the trigger is unpleasant. This, just like punishment, creates fear and distrust, which slows learning. It is very hard for an owner to train a rabbit if the rabbit is scared of them.

How can I resolve a specific behavioural problem?   159

Method 4: Wait for the behaviour to go away on its own

Rabbits perform behaviours to achieve certain goals. The rabbit that bites the owner when the owner puts their hand into the rabbit's hutch has found that this behaviour has a desirable outcome: the owner withdraws their hand. If the rabbit finds that the behaviour stops being effective, then the behaviour will gradually diminish in frequency and will stop. Additionally, reducing the frequency at which a rabbit performs a behaviour (by setting up the environment so this is not necessary, for example) reduces the rabbit's habitual reliance on this behaviour. This is called 'extinction' of the behaviour (Diagram 5.4).

Traditionally, this was the method that was recommended to train a rabbit to be handled. The owners were advised to show the rabbit that the aggressive or fearful behaviour did not have any effect on the owner's behaviour: that they should wear thick gloves to prevent themselves being bitten and persevere in picking up the rabbit until it stopped biting them. Gradually, the rabbit was supposed stop biting the owner.

There are several problems that can occur with Method 4. Failure to consider the rabbit's motivations for performing a behaviour can cause substantial stress to the rabbit and will adversely affect its relationship with its owner. Continuing to handle a fearful rabbit may cause extinction of the behaviour, but may also create a 'learned helplessness' state where the rabbit learns that it has no control over a situation: this adversely affects its welfare. Some behaviours (such as wallpaper stripping) are self-reinforcing or fulfil the rabbit's frustrated behavioural requirements. The rabbit is unlikely to stop performing these behaviours. Additionally, behaviours that are very intermittently reinforced are unlikely to extinguish. If a rabbit bites a glove-wearing owner on a non-glove-wearing part of their anatomy, causing the owner's withdrawal, it will be much more likely to continue showing this behaviour in the hope of preventing more unwanted interactions.

There are some instances where behaviours will extinguish on their own. A good example is the behaviours that are shown when rabbits are establishing a social hierarchy. Initially, a pair of bonding rabbits may interact in different ways: there may be chasing, mounting or agonistic interactions. As the rabbits start to bond, they will learn how to get along and will no longer need to use these behaviours, and hence the behaviours will diminish in frequency without any intervention from the owner.

Method 5: Train an incompatible behaviour

This is one of the most effective methods when changing a behaviour in a rabbit (Diagram 5.5). Rabbits learn quickly, and, if on an appropriate diet, they are motivated to work for treats. Is the rabbit getting under the owner's feet in the kitchen? Teach it to go out of the kitchen on cue, and reward it when it is outside. Sitting by the door is incompatible with getting under the owner's feet.

Logic	Example
Rabbit performs unwanted behaviour	Owner does not want rabbit to go into certain room
Owner does not reward behaviour	Owner ignores rabbit when it is in room
Rabbit stops performing behaviour	Rabbit does not find room interesting

Diagram 5.4: Wait for behaviour to extinguish.

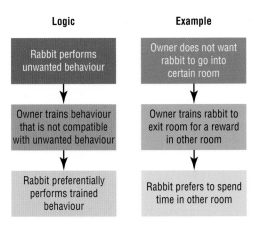

Logic	Example
Rabbit performs unwanted behaviour	Owner does not want rabbit to go into certain room
Owner trains behaviour that is not compatible with unwanted behaviour	Owner trains rabbit to exit room for a reward in other room
Rabbit preferentially performs trained behaviour	Rabbit prefers to spend time in other room

Diagram 5.5: Train incompatible behaviour.

Method 6: Put the unwanted behaviour on cue

In dog training, when a behaviour is brought under stimulus control, or put on cue, it tends to extinguish in situations when the cue isn't given (Diagram 5.6). This tends to have less utility during rabbit training, because most unwanted behaviours are in response to husbandry-related problems, rather than interaction-related problems.

Trying to put a normal behaviour (such as elimination) on cue is hard (the rabbit is

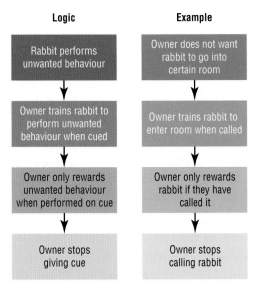

Logic	Example
Rabbit performs unwanted behaviour	Owner does not want rabbit to go into certain room
Owner trains rabbit to perform unwanted behaviour when cued	Owner trains rabbit to enter room when called
Owner only rewards unwanted behaviour when performed on cue	Owner only rewards rabbit if they have called it
Owner stops giving cue	Owner stops calling rabbit

Diagram 5.6: Put unwanted behaviour on cue.

motivated by internal rather than external stimuli, so forming the cognitive link between a cue and the behaviour is difficult), the internal stimuli will override the external stimuli (a full bladder is likely to prompt urination regardless) and, even if it were possible, it would not be desirable to have a rabbit that only urinated when told to do so.

This method could be an option for a rabbit that shows defensive behaviours when the hutch is being cleaned out (including boxing or biting at the brush). When the rabbit shows these behaviours at one specific tool, the behaviour can be rewarded (a suitable reward might be withdrawal of the brush, as this is the goal of the behaviour). For all other tools, the behaviour would be ignored and the cleaning would continue. The rabbit would learn that the behaviour was only effective on one particular tool, so the behaviour is likely to diminish towards other tools. However, there would be more humane ways to change this behaviour – the rabbit is stressed by the invasion of its territory, and this method of behavioural change is not addressing the underlying motivation. A more humane option would be to provide a pile of food out of sight of the hutch and encourage the rabbit to eat elsewhere while the hutch is cleaned.

Method 7: Shape the absence of the behaviour

This method is useful for very fearful or aggressive rabbits, and is the major method involved in constructional approach training. This method reinforces everything that is not the undesired behaviour (Diagram 5.7). So if the rabbit shows a freeze response when the owner approaches, and the owner wants to reduce expression of this behaviour, the owner can move away (this constitutes a reward for a fearful rabbit) when the rabbit shows any behaviour other than stillness: raising the head,

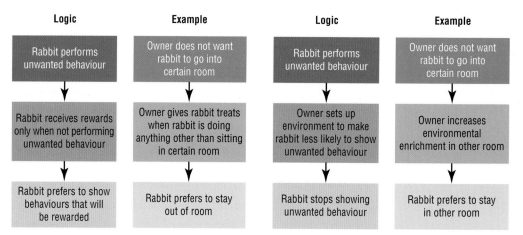

Logic	Example		Logic	Example
Rabbit performs unwanted behaviour	Owner does not want rabbit to go into certain room		Rabbit performs unwanted behaviour	Owner does not want rabbit to go into certain room
Rabbit receives rewards only when not performing unwanted behaviour	Owner gives rabbit treats when rabbit is doing anything other than sitting in certain room		Owner sets up environment to make rabbit less likely to show unwanted behaviour	Owner increases environmental enrichment in other room
Rabbit prefers to show behaviours that will be rewarded	Rabbit prefers to stay out of room		Rabbit stops showing unwanted behaviour	Rabbit prefers to stay in other room

Diagram 5.7: Shape absence of behaviour.   **Diagram 5.8:** Change motivation.

moving the ears, looking in a different direction. In rabbits that show aggressive behaviours in the hutch, the rabbit can be clicked and rewarded for any behaviour other than aggressive behaviours (reward the rabbit when its ears are upright, when it moves forward gently, when it grooms itself).

Method 8: Change the motivation of the rabbit

This is the most humane way to change a behaviour. Rabbits are relatively recently domesticated. They are often kept in a way that does not align well with their motivations. There is usually a lot of scope to improve husbandry, so simple changes can be sufficient to resolve problem behaviours in a way that benefits both the rabbit and the human. This approach causes less stress for the rabbit and has a good chance of success because the owner does not have to put ongoing work into training the rabbit (Diagram 5.8).

Teaching owners to read rabbit behaviour and understand rabbit motivation is very important here. For example, some rabbits find being handled so unpleasant that the owner can't provide a reward that is strong enough to overcome this aversion. In this case, the owner

needs to find alternative ways to manage the rabbit so it does not need to show aggressive behaviour. When an owner stops picking up a rabbit, and starts hand feeding it, its motivation to bite the owner's hands will be greatly reduced and the behaviour will extinguish on its own.

Method 8 is especially useful for inappropriate urination. The owner wants the rabbit to urinate in its litter tray. The rabbit wants a place to urinate that is close, comfortable and has olfactory signals for urination. If a rabbit has a very large area to roam in, it may not get back to its litter tray in time. Once it has urinated in a different place, that olfactory signal will encourage it to urinate there again. The owner can change the motivations here in a number of ways: by providing litter trays in different places around the rabbit's territory, by increasing the desirability of the litter tray as a place to spend time (putting fresh hay in it to encourage the rabbit to eat there) and by using a biological cleaner to remove traces of urine scent to decrease the motivation to urinate in the undesired location.

Table 5.1 shows how each method could be used to change three exemplar unwanted behaviours (the first example has already been described individually for each method).

Table 5.1 Options for changing unwanted behaviours.

Options for changing the behaviour	Preventing a rabbit going into a room	Preventing aggressive behaviour	Preventing the rabbit from urinating on a chair
Method 1: remove opportunity	Put a gate at the door	Rehome the rabbit	Move the rabbit into an outdoor enclosure
Method 2: punish the behaviour	Throw something at the rabbit if it comes into the room	Tap the rabbit on the nose when it bites you	Throw a cushion at the rabbit when it urinates on the chair
Method 3: negatively reinforce the behaviour	Shout at the rabbit and only stop when it leaves the room	Hold the rabbit while it is biting you, let it go when it tries to move away	Make an unpleasant noise when the rabbit starts urinating, stop when the rabbit stops
Method 4: wait for the behaviour to extinguish	Ignore the rabbit when it comes into the room	Wear thick gloves and persevere with handling	Keep thoroughly cleaning the chair
Method 5: train an incompatible behaviour	Train the rabbit to go into its bed on command	Train the rabbit to take food from your hand	Train the rabbit to hop off the chair on command
Method 6: put the behaviour on cue	Teach the rabbit to come into the room when called. Then don't call it	Teach the rabbit to bite your hand when you give a signal. Then don't give the signal	Train the rabbit to urinate on command. Don't give it the signal unless it is in the right place
Method 7: shape the absence of the behaviour	Give the rabbit treats when it stays out of the room	Give food rewards when the rabbit doesn't show aggressive behaviour	Reward the rabbit when it is not toileting in an inappropriate place
Method 8: change the motivation	Provide more enrichment in the room that you want it to stay in	Learn to read the rabbit's behaviour and stop interactions before the rabbit shows aggressive behaviour	Provide easy access to better places to toilet

Rabbit–human interactions

The previous section discussed general ways of preventing unwanted behaviours. This section considers specific examples of problem behaviours related to the interactions between the rabbit and human, and appropriate ways to resolve them.

Aggression

Aggression in rabbits is common – Welch (2015) reported that it was the most common undesirable behaviour. Aggressive behaviours include:

- threatening: abruptly orientating head toward a group mate with half-closed eyes and mouth ajar

- attacking: abruptly running toward a group mate, invariably followed by other aggressive behaviours
- chasing: pursuing another rabbit that is running away
- biting: biting a group mate, may hold on or draw blood
- fighting: often reciprocal wrestling with biting and repeated kicking with fore and hind limbs.

Female rabbits are much more likely to show owner-directed aggression, stranger-directed aggression, and fear of strangers (d'Ovidio et al., 2016). Most female rabbits showed owner-directed aggression only when held or manipulated, rather than around food or when approached in the cage (so unlikely to be caused by competition over food or territorial defence). There were no differences in the expression of this behaviour between neutered and unneutered females, or between those acquired from different sources: so it is unlikely that this behaviour is hormonal or related to inadequate social exposure during early life.

Rabbits in the wild may show aggressive behaviour for a number of different reasons. In a wild social colony, aggressive interactions among female rabbits are aimed to acquire and defend exclusive access to warrens. In male rabbits, the function of aggression is thought to be to establish dominance and mating rights. However, in a domestic environment, the rabbit may show aggressive behaviour towards an owner when competing for food, when defending a perceived territory, when fearful or when frustrated.

In an established social hierarchy in a wild rabbit colony, these behaviours are rarely seen – the rabbits signal to defuse tension so they don't waste energy fighting or risk being injured. If one rabbit shows aggressive behaviour to another, the response of the attacked

rabbit may be to move away from the attacker, or to adopt a submissive posture (laying back of the ears, sometimes flattening the head, forepaws or whole body to the ground). Both of these reduce the arousal level and the expression of aggressive behaviour from the other rabbit.

If an owner reports that their rabbit shows aggressive behaviour towards them, they typically describe lunging, pouncing, scratching or biting. These visible behaviours occur when the rabbit has tried, unsuccessfully, to use more subtle behaviours to stop the owner's actions. The owner needs to learn to read the escalating signs of aggression and stop their interaction before the rabbit inflicts injury. If they merely try to reduce the incidence of the most visible behaviours, the rabbit will still be stressed, and the owner may not have much success. The most humane way to resolve aggressive behaviours is to decrease the rabbit's motivation to show any of these behaviours.

If a rabbit shows aggressive behaviour, it is stressed and is experiencing poor emotional welfare. These behaviours may both cause, and be caused by, a poor human–animal bond, which may adversely affect other aspects of the rabbit's welfare.

When dealing with any form of aggression, the owners need to learn the behavioural signs that indicate increasing stress, or situational triggers that cause stress, and defuse the situation. The owners may increase their rabbit's fear and distress by either persevering with an action that the rabbit finds unpleasant, or by punishing the rabbit's behavioural expression of stress.

There has been much advice written about addressing aggressive behaviours in rabbits, but some of these recommendations are unlikely to be successful (see table 5.2).

Table 5.2 There is a lot of 'advice' available on how to reduce aggressive behaviour in rabbits, but much of it is not evidence-based and can actually worsen the situation.

Myth	Reality
'Tap the rabbit on the nose when the rabbit bites you.'	Punishment is usually ineffective in modifying a rabbit's behaviour. This is even more applicable when the rabbit is already fearful or stressed: punishment will increase these emotions and is likely to worsen the aggressive behaviour.
'Make a high-pitched squealing noise when the rabbit bites you.'	Rabbits make a high-pitched squealing noise when caught by a predator, so this noise is associated with danger. If a rabbit bites deliberately to cause the owner to withdraw, it already feels fearful or distressed, so this strategy is likely to increase these emotions and worsen the aggressive behaviour. Sometimes, it works once or twice (as it distracts the rabbit), but then rapidly stops working as the rabbit desensitises to it.
'Wear gloves when handling your rabbit, so it learns that biting you doesn't make you withdraw your hands.'	This may reduce the rabbit's expression of the biting behaviour, but at the cost of increasing its stress levels: the owner is still doing something to the rabbit that it intensely dislikes. When an owner perseveres with an interaction that causes fear or distress to the rabbit, they are failing to respect the rabbit's needs.
'Train yourself not to flinch when the rabbit bites you.'	This is not fair on the owner and will not work!

Owners should be educated in rabbit social behaviour and encouraged to respect the rabbit's wants and needs by withdrawing earlier when the interaction is not welcome. This will build the rabbit's confidence and trust in the owner, and decrease the incidence of aggressive behaviours.

Bolder rabbits are more likely to show overt aggression when scared; timid rabbits are more likely to withdraw. If the environment is appropriately set up and the interactions are on the rabbit's terms, 'aggressive' rabbits can make extremely good pets that are highly trainable and confidently seek out human contact.

Case study 6: Babbitt

Babbitt was rehomed because she was 'aggressive'. She would lunge, growl and scratch if people tried to pick her up. These aggressive behaviours stopped when she was no longer picked up. However, throughout the rest of her life, when people approached, she would run directly towards them with her ears clamped down – typically a show of aggression.

However, Babbitt showed this behaviour when she wanted to be stroked. She was an extremely confident rabbit, which probably increased the likelihood of showing aggressive behaviours when her welfare was poor, but she became very friendly and interactive when her welfare needs were met.

There are several different types of aggression, and these are grouped according to the type of behavioural modification that is effective. These categories are territorial, fear- or frustration-related and food-related.

Territorial aggression

Some rabbits show aggressive behaviours when the owner approaches their cage or tries to clean it out. This not only damages the relationship between the human and rabbit, but this territorial aggressive behaviour often reduces the frequency at which the owner cleans the cage. This may affect the rabbit's physical health.

Territorial aggression commonly presents as aggressive behaviour when the owner touches objects within the hutch (food bowl, litter trays, etc.). There are various husbandry practices that increase the risk of territorial aggression. It is, for example, more common in singly housed rabbits because such rabbits are dependent on their environment, rather than their companion, for emotional stability. It is more common in rabbits confined in small areas: in larger territories, the rabbit has more access to resources (so there is less need to defend any one resource) and can more easily retreat or move elsewhere.

To reduce territorial aggression, these husbandry factors should be rectified. Singly housed rabbits should be bonded to a companion, and the environment should be assessed and improved where possible.

When environmental modification is not possible, or is only partially effective, there are two methods of reducing the incidence of territorial aggression – desensitisation and counter-conditioning, and distraction.

Desensitisation and counter-conditioning

It may be possible to desensitise the rabbit to the owner moving items around in the cage by gradually increasing the intensity of the stimulus (initially moving close to objects, then touching them, then moving them) and providing a reward when the rabbit does not show aggressive behaviours. Although this seems to be a nice 'complete' solution, there are various reasons why this is often ineffective.

First, during a desensitisation programme, the animal should never be exposed to stimuli of a higher intensity than it can cope with – however, the owner needs to be able to clean out the cage frequently. Second, in many rabbits, the aversive experience of an owner changing their territory is greater than the pleasurable experience of a food reward, so the counter-conditioning is hard to achieve. Third, although the aim is to desensitise the rabbit, some animals can become sensitised, so the problem will worsen (see section on Non-associative learning at p. 133 for a discussion of sensitisation).

Distraction

This method involves helping the rabbit to 'ignore' the intrusion on to their territory by reinforcing another behaviour in a different location. The first step is to teach the rabbit to recall or to target an object, so it can be moved out of the cage on command. The principle of this method is that the rabbit receives a reward for remaining outside the hutch, and the rabbit cannot see that the owner is cleaning the hutch. Breaking the line of sight of the rabbit (using a barrier) shifts the balance of attention between the food reward and the cage cleaning – it is easier for the rabbit to 'ignore' the territorial intrusion, and therefore the food reward is likely to be sufficient. Food rewards should take a while to consume: a pile of grass or leafy vegetables is suitable.

The owner should start by calling the rabbit over to them (preferably to an area that can easily be blocked off from the area to be cleaned). The owner should provide a pile of grass or vegetables. While the rabbit is feeding, the owner should put up a barrier (such as cardboard boxes) while they clean out the area.

Once the owner has finished, they should step over the barrier, give a higher value food reward from the hand and then remove the barrier.

The aim is for the rabbit to learn that the owner's presence near the cage doesn't cause a noticeable intrusion on to its territory. When the rabbit is relaxed with the owner near the cage, the owner should use the recall command to move the rabbit into the cage for a reward. The rabbit then learns that the owner's presence near the cage can be rewarding.

Case study 7: Bella

Bella was a neutered female rabbit awaiting rehoming at a rescue centre. Her enclosure was large and well designed. However, she started to show territorial behaviour around the hutch area that contained her litter trays. She started by circling and running at the staff who were trying to clean these out, then started to growl, lunge, and bite their feet.

The staff changed their routine for managing Bella – before cleaning the trays, they put palatable food in an area of the enclosure that was partially out of sight of the hutch, and created a makeshift visual barrier to reduce her stress levels while they were cleaning. This provided sufficient distraction that Bella could ignore the repeated invasion of her territory as she was more motivated to eat the food than she was to defend her territory.

Fear- or frustration-related aggression

These two motivations are grouped together because the presentation and treatment of these behaviours is similar, and differentiating the emotional state of the rabbit may be difficult. Typically, the owner reports that the rabbit shows aggressive behaviours when the owner tries to interact with it. Most owners who report aggression in their rabbits frequently pick up their rabbits (either to show affection or for practical reasons) – this restraint may cause either fear or frustration. To reduce the incidence of aggressive behaviours, the stimulus for the aggressive behaviour must be removed: either permanently (where this is possible), or at least temporarily for the purposes of desensitisation and counter-conditioning.

Owners should be educated about positive interactions with their rabbit (giving the rabbit choice, mutually rewarding interactions), and about those interactions that the rabbit perceives as unpleasant (being picked up). Owners should be asked how they 'play' with their rabbits: many owners mistake frustration behaviours for play behaviours, which will increase the risk of their rabbit showing frustration-related aggression.

The aim of a behaviour modification plan to reduce fear- or frustration-related aggressive behaviour is to ensure that the rabbit never needs to show aggression (as shown in section on Resolving an unwanted behaviour at p. 158, reducing the frequency at which the rabbit performs the behaviour reduces its habitual reliance on the behaviour). If the owner knows the trigger that causes aggression, then the trigger should not be given (the owner should alter the rabbit's environment or change their own behaviour). If the owner does not know the trigger, they should learn the signs of stress in the rabbit (see Diagram 5.9) and stop interacting with the rabbit when it shows these signs. If the owners have previously ignored signs of fear or frustration, the rabbit learns that it needs to escalate its behaviour to achieve the desired outcome (the owner withdrawing). Once the owner starts to retreat at subtle signs of stress,

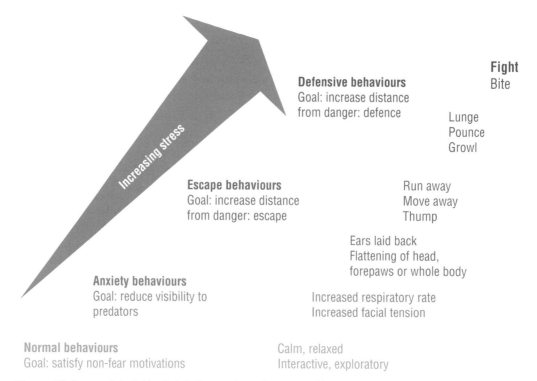

Diagram 5.9: Stress escalation ladder: the behaviours escalate as the intensity of the stress increases.

the rabbit learns that it does not need to show the higher-energy aggressive behaviours.

Once the owner has stopped their aversive interactions with the rabbit (such as picking up the rabbit, interacting with the rabbit when it doesn't want to), he or she must start to increase the rewarding interactions with the rabbit. The owners should start to hand feed concentrate food or treats: this increases the rabbit's willingness to approach the owner and provides a foundation for training a few behaviours. The owner should also learn to 'ask' the rabbit if it wants to be stroked (see Advice sheet 12), and respect the rabbit's choice. In general, if a rabbit can predict or influence its owner's behaviour, it will be less likely to use fear- or frustration-related aggression to cope.

Food-related aggression

Food-related aggression is relatively common in pet rabbits. It is caused by the restricted provision of high-value food items in a small area – wild rabbits, with access to large areas where most resources are widely distributed, almost never show this behaviour. This type of aggressive behaviour is directed at perceived 'competitors' for food resources, either the owner or the companion rabbit (Figure 5.1). Owners typically report that when they go to put food in the rabbit's bowl, or take the bowl out, the rabbit growls or bites them. Moving the rabbit on to a more natural feeding regime will reduce the incidence of this behaviour.

If an owner reports food-related aggression, then the diet of the rabbit should be assessed, both in quality and quantity. The amount of concentrate food should be reduced or preferably eliminated. If the owner wants to continue feeding concentrate food daily, the food should not be put in a bowl, which creates a focus of attention, but should be scattered in the hay. This stimulates foraging behaviour and prevents

Figure 5.1: Bowl feeding groups of rabbits increases the risk of food aggression. Owners should be advised to scatter food rather than provide a resource focus that can be defended.

the rabbits from defending a resource focus. In addition, the owner should decrease the predictability of the pellet 'meals' – differing amounts at different frequencies and different times of day.

Another solution is to do some basic training, giving concentrate food as rewards for desired behaviours (such as recall or targeting) rather than rewarding unwanted behaviours (when rabbits lunge at their owner, the owner drops the food into the bowl more quickly). Encouraging the owner to move to hand feeding concentrates solely as rewards for desired behaviours will improve the rabbit's behaviour and improve the owner's relationship with their rabbit.

Case study 8: Nutmeg

Nutmeg was a one-year-old Rex female rabbit. She lived happily with a neutered male companion, and the owner had no problems except at feeding time when Nutmeg rushed at the bowl, growling and biting. The food was either pellets or leafy vegetables. Outside of feeding times, the owner could hand feed her treats, which she ate well. The owner wanted to have Nutmeg neutered, because she thought that would sort out the problem.

The owner was advised to wait to spay Nutmeg until the behaviour was resolved, as the disruption caused by the surgery would delay her learning. The owner changed the way that she fed Nutmeg and her companion. When she fed leafy vegetables, she would call both rabbits over and give them each a piece to eat while she put the rest in their cage. She increased the amount of hay provided, reduced the amount of concentrate food, which she sprinkled into the hay, and removed the bowl. Nutmeg's behaviour changed within a week, and the owner was very pleased with the outcome.

Fear-related problems

Fear-related problems may present in a variety of ways. Owners who can read their rabbit's behaviour may report that the rabbit doesn't like them – either that it retreats from attempts to interact with it, or that it never tries to interact with them. This can be very frustrating, especially for highly motivated owners who try very hard to provide their rabbit with everything it needs. Owners who find it harder to read their rabbit's behaviour may report that the rabbit repeatedly thumps its feet when they try to interact with it. Rabbits that thump their feet repeatedly when the owners are not trying to interact with them may be fearful of something else (exposure to other predator species, for example), or may be showing frustration with some aspect of their environment.

Timidity and disinclination to engage arise from mismatches between the owner's behaviour and the rabbit's emotional needs. Most commonly, owners who complain of fear-related behaviours frequently pick up their rabbits, or they may also be interacting with the rabbit in another way that causes fear or distress. Occasionally, however, disinclination to engage may occur because the rabbit has an excellent environment that meets its requirements, and the owner is not a source of interest or reward in the rabbit's life.

Case study 9: Panda and Oreo

Panda and Oreo were a bonded pair of neutered one-year-old English rabbits.

They had been bought as pets for a young family. The owners had done some reading before acquiring the rabbits – they lived outside, had a large enclosure with plenty of toys and were fed on a good diet. The rabbits had been acquired at eight weeks of age. Initially, the children were able to interact with them and handle them, but the rabbits progressively got more flighty and the children were no longer able to get near them.

To improve the relationship between the owners and rabbits, the owners were taught about the motivations of the rabbits, and they passed this on to the children. The children were encouraged to spend time sitting in the rabbits' enclosure without moving towards them, and were encouraged to feed them pieces of vegetable and concentrate food. As the rabbits got more confident, the children could use food rewards to encourage the rabbits to climb on to them and to jump over toys: this made the rabbits much more interesting to the children, and made the children less frightening to the rabbits.

The parents watched their children teach visiting friends about interacting with their rabbits, seeing the situation from the rabbits' point of view. The owners were pleased at the change in the rabbits' behaviour, and also commented that it had taught the children some impulse control!

If the rabbit is fearful of being picked up, then the owner should be taught to see the interaction from their rabbit's point of view. Humans are unusual among mammals in that they explore the world with their hands, rather than with their nose and face. Human babies enjoy being picked up and cuddled, so humans think that other animals will also enjoy this. As rabbits don't have hands, they have no positive associations with being picked up. Most animals do not enjoy being picked up by an animal of a different species, especially if the other animal is a predator. Rabbits that are frequently picked up are much less likely to voluntarily approach the owner.

Advice on setting up the environment to avoid the need to lift a rabbit can be found in section on Providing a more stimulating environment at p. 130.

In some situations, owners may be doing something else that the rabbit finds aversive. Owners should be asked whether they 'trance' or 'hypnotise' their rabbit: this is extremely stressful and should not be performed by owners (although it can be a useful diagnostic test: if owners are successfully able to trance their rabbit, this confirms that their rabbit is fearful of them, since rabbits can only be successfully tranced if they are scared). Ask the owners how they stroke their rabbit: rabbits prefer to be stroked on the face than on the back, and very much dislike having their abdomen or feet touched. Additionally, try to find out whether the owner tries to inappropriately 'play' with their rabbit: chasing or teasing the rabbit can result in frustration and stress.

If the owner does not frequently pick up their rabbit, and the environment is good, then the owner should be encouraged to interact with the rabbit in a way that it finds more interesting. Hand feeding food is a good way to increase approach and investigative behaviours, and provides a foundation for the owner to begin training their rabbit (see next section). Once the rabbit learns that the owner is a source of reward, it will be more interested in engaging with them.

Difficulty catching rabbit

As discussed in the previous section, rabbits dislike being picked up and restrained: being 'caught' is very unpleasant. Owners should follow the recommendations above to reduce the need to catch the rabbit.

However, if the rabbit does escape, it will need to be caught. Owners should teach the rabbit to obey a recall command: this is useful to try as a first attempt to bring the rabbit back to a place of safety if they escape. Additionally, an owner who very rarely picks up their rabbit will find it easier to catch the rabbit if they need to as the rabbit won't anticipate being caught. Training a recall command is covered in Advice sheet 8 on clicker training.

Sexual or reproductive behaviour directed at people/objects

Male sexual behaviour

Sexual behaviour directed at humans or objects is seen most commonly in singly housed unneutered male rabbits (although by no means in all unneutered male rabbits). Male rabbits become fertile at about three and a half months old, although may start showing sexual behaviour earlier. Male rabbits 'grunt' or 'honk' to signal sexual intent, and may spray the owner with urine. The rabbit may then attempt to mount objects or the owner's hands or feet, grasping the owner's skin in its teeth.

This behaviour is extremely uncommon in group-housed rabbits: the male rabbit has a more appropriate target for this behaviour. There are two goals for treating this behaviour in the rabbit: to reduce the hormonal drive to perform the behaviour, and to provide social companionship, as expression of this behaviour indicates that the rabbit is likely to be suffering from lack of species-specific companionship. This can be a difficult message to convey, as neutering alone will diminish incidence of the behaviour, so owners may not see why they should acquire a second rabbit. However, merely preventing the hormonal incentive to perform the behaviour does not result in better welfare for a rabbit kept in social isolation.

Owners usually are more concerned by mounting behaviour directed at them rather than directed at objects. This is a common reason for owners to request that their rabbit is castrated. This behaviour is primarily hormone related and should diminish over the four to six weeks after removal of the testes. This intervention provides an opportunity to recommend that the owner introduce a female rabbit to their male rabbit. At this time, their male rabbit will be neutered (so introductions are usually easier), and this unwanted behaviour usually indicates that the rabbit is suffering from lack of companionship.

Female reproductive behaviour

Female reproductive behaviour is not just sexual in origin, but may be related to hormonal changes during pregnancy (or pseudopregnancy) or during nursing of young. Females may start to be receptive to mating at about 12 weeks of age, but will rarely ovulate – they usually reach puberty at 70–75 per cent of their adult weight (around 16 weeks, [Masoud et al., 1986]). Ovulation is triggered by sexual intercourse, and female rabbits are receptive to mating for about 14 days in every 16. False pregnancy (hormonal changes similar to pregnancy but without fertilisation of the eggs) is common in rabbits, and usually lasts 15–18 days. The gestation period lasts 31–33 days.

Shortly before giving birth (or 'kindling'), the female pulls fur from her body and builds a nest for the young rabbits. Litter size ranges between 3 and 12. The female feeds the litter once every 24 hours, and may not visit the nest box in between feeds (a useful adaptation for a prey species that must spend long periods feeding away from the nest [Lebas et al., 1997]). In the wild, young rabbits stay in the burrow for 18 days and are weaned at 21–25 days.

One study found that female rabbits were said to be more 'temperamental', less likely to be described as 'quiet and placid' and more likely to exhibit negative personality traits (Mullan and Main, 2007). Female rabbits are significantly more likely to show owner-directed aggression, especially if kept alone (d'Ovidio et al., 2016). However, the two studies found differing effects on behaviour when females were neutered: the former found that neutered rabbits were more likely to be described as 'quiet and placid', but the latter found no difference between rates of owner-directed aggression between neutered and unneutered females. Most aggressive behaviour in female rabbits does not seem to be related to territory or early life socialisation, but is shown when the rabbits are handled. Neutering female rabbits may not alter aggression – owners should instead understand the motivation for the behaviour and interact with the rabbit in a more appropriate way.

At puberty, female rabbits may start to pluck their fur and spray urine when they are receptive to mating. Urine spraying is much less common in female rabbits than in males. Neutering should decrease the incidence of these behaviours.

Reproductive behaviours in female rabbits occur more frequently when the days start to lengthen in the spring, but usually settle down after a month or so. Neutering does not fully prevent these seasonal changes.

Rabbit–rabbit interactions

We've now explored rabbit–human interactions. This section explores rabbit–rabbit interactions.

Aggression

Aggressive behaviours are hostile behaviours that often lead to injury. Aggression between rabbits can have very severe consequences. Owners may report that the rabbits chase each other, kick each other, pull out each other's fur or bite each other. Once a pair is bonded, then aggression resulting in even minor injury is very uncommon. Even in well-bonded animals, there are occasionally agonistic behaviours from one rabbit to the other, but one animal should defer or retreat from the other to defuse the situation. If it does not, the behaviours will escalate and one rabbit is likely to be injured.

When taking the history, try to ascertain when the aggressive behaviour started, whether there is an obvious cause and whether the environment results in very restricted access to a favoured resource (food bowl, water bowl, hay rack, ledge).

Some owners may perceive normal intermittent agonistic behaviour as aggressive, so video evidence is useful here. Illness or pain may affect the bond between rabbits. A detailed clinical examination of both rabbits is imperative. An ill rabbit may behave in a variety of ways – they are not always the attacked or the attacker.

Rabbits may show aggression towards their companion if the bonding is inadequate, there is competition for a resource (food or territory) or if there is a change in the relationship between the rabbits (frustration, stress or illness). If there is an obvious cause, reduce the constraint (increase access to these resources). If there is no obvious cause, consider medical problems that may be causing this behaviour. If the problem has been present since the rabbits have been together, the bonding process may be incomplete.

If one rabbit is at risk of injury, the rabbits should not be left in the current situation. The rabbits may be separated by a wire fence (so they can see, hear and smell each other), and gradually reintroduced to a changed environment (this can be achieved by moving objects around, reducing lines of sight, increasing access to food) as though they were being re-bonded. This relationship breakdown is uncommon, but usually associated with poor health in one of the rabbits. If they cannot be re-bonded, then they may have to be permanently separated by a wire fence. It is very rare that the cause of the problem cannot be rectified, so this permanent separation is rarely necessary.

Territorial aggression

When one rabbit defends a key resource or territory from its companion, this is termed territorial aggression. In the wild, rabbits live in small groups with several bonded pairs within a larger warren, but show territorial aggression towards animals that do not live in their warren. In captivity, rabbits will still defend their territory against perceived intruders. This creates a tension: rabbit welfare is much better when they are kept with other rabbits, but the introduction period can be very difficult due to the territorial aggressive behaviour.

Territorial aggression should not occur in a bonded pair of rabbits. One of the evolutionary advantages of living in social groups is that the members of the group benefit from proximity of other members. By definition, a bonded pair of rabbits should not display territorial aggression unless one rabbit is defending a very constrained resource or there are frustrated sexual motivations.

Young rabbits that have been housed together since birth may show territorial aggression when they reach puberty. Entire males are likely to fight viciously (though some pairs will be stable, but do not count on this) and females may start to defend a hutch area. To achieve stable relationships between rabbits kept in groups, neutering is usually advisable. If an owner presents a pair of rabbits that are showing territorial aggression and that are not neutered, then neutering should be the first intervention that is tried. Both rabbits should be neutered together and kept together at the veterinary surgery. This stress from an unfamiliar location will help to re-establish the bond.

Be aware that rabbits can inflict serious injuries on each other. Rabbits showing territorial aggression should never be left to 'fight it out' – they should be separated, any cause of the behaviour resolved, and then they should be gradually reintroduced as though they were being re-bonded.

Case study 10: Kirk, Spock and Bones

Kirk, Spock and Bones were three crossbreed male rabbits, from the same litter, bought from a pet shop at eight weeks of age. The rabbits got on well until puberty. When the rabbits were about four months old, the owner came home to find that there had been a severe fight: Kirk and Spock were severely injured.

The owner rushed the rabbits to the vet. Kirk had been eviscerated and had to be euthanised. Spock had sustained severe damage to a hind limb, which was amputated. The owner opted to rehome Spock and keep Bones.

Food-related aggression

Most of the principles described in the section on 'Food-related aggression' towards humans are equally applicable when the same behaviours are shown towards a companion rabbit. In the wild, rabbits rarely need to guard food, as it grows from the ground and is distributed over a wide area.

Owners may report different presentations of this behaviour: aggressive behaviour when a food bowl is placed in the cage, or aggressive behaviour when they give a large food item (such as a piece of cabbage leaf) to one of the rabbits.

Owners should stop feeding the rabbits from a bowl (which creates an artificial resource focus to defend). They should reduce the quantity of concentrate fed to encourage the rabbits to eat more forage. Forage is better for their physical health and will increase the time they spend feeding. If the owner wants to continue giving concentrate food, they should either hand feed it (using one hand for each rabbit), or scatter it in the bedding.

If the owners want to give larger pieces of food (such as pieces of cabbage leaf), which don't fit into the mouth whole, they should give a piece to the dominant rabbit first before feeding the other rabbit. Usually, the rabbits will carry their piece of food in opposite directions. Owners should give the larger rabbit (or the rabbit that eats faster) a larger piece to try to ensure that both rabbits finish at the same time. This is contrary to the human desire for 'fairness', but when humans try to redress the balance and disrupt the hierarchy between the rabbits, they may inadvertently cause the subordinate rabbit to be harassed by the dominant one.

Displaced aggression

Frustration of any sort can manifest as aggressive behaviour towards a companion.

A common situation is that the rabbit shows aggressive behaviour towards the owner when they remove the food bowl from the cage, and once the owner's hand is out, the rabbit then growls or chases its companion.

To resolve displaced aggression, the triggers of the behaviour should be examined. Does it occur after the rabbit has been handled? Does it occur when the rabbits are restrained in a small space? Is it linked to the provision of food? Once the triggers are identified, they can be eliminated to reduce the frustration.

Sexual behaviours

Most owners do not want to watch their pets engaging in sexual behaviours. Male–female rabbit pairs will reproduce if not neutered, but even if they are neutered, sexual behaviours are still relatively common – during bonding, in spring and as part of normal rabbit interactions.

Mounting frequently occurs during the bonding process, and fur is often pulled out from the back of the rabbit being mounted. If it has space to do so, the rabbit being mounted will move away when it wants to stop the behaviour. Owners should not try to intervene unless it is causing a fight, (where neither rabbit is moving away and they are trying to injure each other with paws or teeth).

Even in well-bonded pairs, mounting is not uncommon. Both male and female rabbits may mount their partner, and may mount either the front or back end. This occurs more often during spring, because the hormonal drive for this behaviour retains a seasonal element even if the sex organs have been removed.

This is entirely normal behaviour. Some owners think that this behaviour will not occur in a neutered rabbit, and so may be concerned that the neutering procedure was not successful. When a male rabbit is castrated, it is extremely difficult to leave testicular tissue behind. It is possible to leave ovarian tissue remnants in a female rabbit. Owners should be alert for the behavioural signs of a false pregnancy (pulling fur from dewlap and abdomen, mammary development, nest building and territorial aggression around the nest). However, mounting alone is not usually associated with ovarian tissue remnants. Owners should be reassured that this behaviour does not mean that the neutering procedure was not successful. If they want to stop the behaviour, advise them to distract the rabbits by scattering food rather than trying to stop the behaviour by punishing the animals.

A significant change in the environment (for example, moving free-range rabbits into a restricted run environment) can trigger mounting behaviour because of a disrupted bond, boredom, or displaced frustration behaviour. This should settle down over time, but such behaviour in this context is a sign of a decrease in the rabbits' welfare.

Case study 11: Teddy

Teddy was a three-year-old Netherland Dwarf, who had been kept singly since he was ten weeks old. The owners wanted to get him a companion, and sought veterinary advice because the rabbit had never shown any form of sexual behaviour or urine spraying. Having read up on the subject, they knew that, when introducing rabbits, the male should be neutered, but they wondered if this invasive procedure was necessary in a rabbit that they considered to be asexual.

The vet advised that Teddy should be neutered. He was introduced to his new companion Velvet six weeks later (she had been neutered six months previously). The owners were shocked that he showed a lot of mounting behaviour towards Velvet, which gradually resolved. Thereafter, each spring, he would show more mounting behaviour, and this was also expressed if the two rabbits were confined in a small space. Four years later, Velvet had to be euthanised because of a mammary tumour, and Teddy was bonded to a new companion. He again showed significant mounting behaviour with the new rabbit.

Barbering

Barbering describes the behaviour where a rabbit pulls or chews the hair of another rabbit. It has been reported in laboratory rabbits and is a rare behavioural problem that arises from some form of stress (Fox, 2012). The 'barbered' rabbit usually presents with patches of broken-off hairs along the neck, legs, chest and lateral flanks – often in places that the barbered rabbit can't see.

When hair loss is suspected to have a behavioural cause, it is important to ascertain whether the damage is being done by the rabbit itself or by its companion. Entire female rabbits will pull fur from themselves (often from the dewlap, ventral abdomen and forelegs) to build nests. This should be ruled out as a possible cause.

If barbering is suspected, ask the owner to provide videos of the two rabbits interacting at other times – this will provide information about the degree of 'normal' interactions and the degree of barbering. The treatment depends on the frequency of the behaviour and its consequences.

If there are obvious problems with the husbandry, the behaviour is being performed infrequently and the barbered rabbit is not suffering,

then the rabbits do not need to be separated. There are several possible causes of this problem behaviour. First, rabbits that cannot express normal behaviours may show abnormal ones such as barbering. Any husbandry problems should be rectified to reduce frustration (increasing space to allow the barbered rabbit to move away; providing more opportunities for the rabbits to express normal behaviours). Second, rabbits that do not eat sufficient long forage are more likely to chew inappropriate substrates that resemble grass – such as rabbit fur. The rabbits should be transitioned on to a grass- or hay-based diet.

If the barbered rabbit is being injured, or if the barbering behaviour is compulsively and frequently performed, then the rabbits may need to be physically separated for a period to break the habitual interaction and allow any injuries to heal. If possible, the rabbits should remain in visual contact and physical contact through a mesh to reduce the stress of separation. This may be enough to prevent the behaviour, but owners should be aware that a determined rabbit could still barber its companion through mesh. The environment should be improved and the diet addressed as previously described. The owner can try reintroducing the rabbits after a couple of weeks, when the hair has mostly regrown.

If the barbering behaviour is habitual, it will be hard to break. Reducing the time budget that the rabbit can spend chewing its companion's fur may reduce this behaviour. Pragmatically, if the behaviour continues but the barbered rabbit's welfare is not affected (no injury, many affiliative interactions, no stress), then this outcome is not a problem. However, if the barbered rabbit's welfare is suffering, then the rabbits may need to be permanently separated, but this is the least desirable outcome.

Conspecific overgrooming

Allogrooming, as described in section on Relationships between rabbits at p. 74, is a common affiliative behaviour between bonded pairs of rabbits. This grooming is usually focused on the face: one rabbit will lick the other on top of the head, around the eyes and along the ears.

The degree to which rabbits groom each other varies both between different pairs of rabbits, and between the same pair in different situations. During times of stress (such as initially after bonding or when the environment is changed), the level of social grooming can be substantially higher. Some owners report that female rabbits start to overgroom their companions after being spayed. This may be due to an increased reliance on the companion rabbit for emotional security during times of disruption. When the higher level of grooming results in unintended injury to the rabbit being groomed, this is called conspecific overgrooming (Bradbury, 2016).

Case study 12: Sooty and Sweep

Sooty (female) and Sweep (male) were a pair of bonded Lionhead rabbits. The owner sought help because Sweep had developed lesions around his eyes. The owner reported that Sooty groomed Sweep much more than Sweep groomed Sooty, and she tended to focus grooming around his eyes. Sooty had never shown any aggressive behaviour towards Sweep and he appeared to enjoy being groomed.

Several weeks previously, the owner had noticed a small hairless patch under Sweep's eye, but the hair had grown back. The owner then noticed a larger hairless patch, which Sooty continued to groom until the skin was raw. Sweep still didn't move away when she licked this area. His eyes were healthy and he showed no other clinical signs.

The owner was very reluctant to separate the rabbits as they had free run of the house, and they had previously been very distressed when separated because Sooty was being spayed. The owner was advised to use the honey method as described here. Sooty immediately redirected her grooming behaviour to the top of Sweep's head, and the owner reduced the honey application as the wound scabbed and healed. The rabbits' husbandry was already very good, but the owner additionally started to use puzzle feeding techniques.

Typically, injuries present as swollen, hairless reddened patches, but there may be open sores if the outer skin barrier is disrupted. The grooming rabbit does not use its teeth – the injury forms when the area is licked repeatedly. If an injury does develop, the change in texture (and presumably taste) may increase the focus of the grooming rabbit on that area, creating a positive feedback cycle and preventing healing.

The pattern of injury differs substantially between barbering and conspecific overgrooming. In barbering, the hairless patches are usually around the underside of the neck, legs and chest, whereas in the latter, the patches are more commonly around the face, and especially the eyelids, where the skin and fur covering are thin.

Management of conspecific overgrooming requires identification of the inciting cause (if possible) and management of this stress. Unlike with barbering, it is important not to separate the rabbits: this will cause further stress, may damage the bonded relationship between the rabbits and will increase the likelihood of over-grooming when the rabbits are reintroduced. Advising the owner to wait and see if the problem resolves may also be counterproductive – when one rabbit grooms the other, the different texture and taste of the wound may reinforce the licking behaviour. Physically preventing the rabbit from licking its companion will be difficult and will result in poor welfare. Elizabethan collars (also called Buster collars or pet cones) prevent caecotrophy and should be avoided in rabbits.

There are two aims when treating conspecific overgrooming – to allow the injury to heal, and to prevent the rabbit overgrooming its companion in the future. Redirection of the allogrooming behaviour away from the patch is often successful. Honey can be used to provide a taste reward for an alternative behaviour – moving the grooming away from the site of injury to an area that has thicker fur, so is less prone to damage. Owners should apply a small smear of very dilute honey between the eyes of the overgroomed animal. This should be repeated on a tapering schedule to break the habit and allow the wound to heal – twice daily for three days and then once daily for a week.

Rabbits in small, barren environments will have a larger time budget to spend on grooming each other. Therefore, the general principles of reducing stress by improving husbandry (described in section on Changing behaviour through changing environment at p. 120) are applicable here – improving diet, providing more space and increasing environmental enrichment can all help to reduce the risk of recurrence.

Rabbit–environment interactions

Elimination

The term 'litter training' was initially applied to cats, and refers to an animal that only eliminates urine or faeces in a litter tray when in the house. A litter-trained cat will almost never urinate or defecate outside of the litter tray. Litter-trained rabbits, by contrast, may continue to deposit some faecal pellets outside the litter tray – this is normal behaviour. Owners may have unreasonable expectations of how 'litter-trained' rabbits should behave. Owners should be educated that litter-trained rabbits should deposit all urine in the litter tray, but may continue to leave some faecal pellets outside it.

In the wild, rabbits mostly eliminate in one area, called a latrine. The majority of urine is deposited at the latrine; some is also used to mark territory by spraying. Faecal pellets are deposited at the latrine, but some are also used to mark a wider territory: these pellets often have a stronger, more 'rabbity' smell. Many rabbits graze while they deposit faeces.

Unwanted elimination behaviours, where the rabbits toilet in a location that is inconvenient for the owner, can occur for several reasons. Some of these reasons are related to pain or disease, some are related to sexual behaviours and some are territorial. Owners are much more likely to be concerned about elimination behaviours if the rabbits live inside and they expect them to use a litter tray.

Owners may describe any form of soiling as a problem with litter training, but the motivations are very different. The first distinction is whether the rabbits are producing urine, faeces or caecotrophs.

Urine

There are two ways that rabbits deposit urine: either through squatting (to eliminate), or through spraying (to mark). The behavioural positions are very different. When a rabbit eliminates, it moves its hind feet forward relative to its body, protrudes its tail and then urinates in a vertical or near-vertical stream, which becomes more pulsatile as the rabbit finishes emptying its bladder. When a rabbit sprays urine, it will circle or run past the object to be marked, rotate its pelvis by 90 degrees and spray urine horizontally.

Inappropriate urination

Rabbits should not eliminate urine (excluding spraying) anywhere other than the litter tray. Urination elsewhere indicates a failure to learn, or a breakdown in the learned behaviour of toileting in the appropriate place. Always check if the rabbit has previously been successfully litter trained: if the behaviour has not been appropriately trained, then the owner needs to perform the process described in Advice sheet 10 in the Appendix.

There are many reasons for a breakdown in litter training.

The rabbit may be:

◆ unaware
◆ painful
◆ fearful
◆ territorial
◆ frustrated
◆ seeking social interaction.

We'll explore these in a bit more detail.

Unaware

The rabbit may be unaware that it is toileting outside the litter tray. This may occur if the litter tray design is inadequate (low sides mean that the rabbit urinates over the side of the tray), or if the rabbit is unaware when it is urinating (i.e. it is incontinent). In the former case, normal volumes of urine will be found next to the litter tray. In the latter case, urine may be deposited all over the environment in small quantities, and may 'dribble' from the rabbit. Incontinent rabbits often smell of urine as the rabbit does not posture to urinate correctly, soiling its fur and hind paws. This requires veterinary attention.

Painful

This may cause the rabbit to have difficulty getting into the litter tray, or may affect the position it adopts to express urine. Additionally, if a rabbit has urinary tract disease, the urge to urinate may be frequent, sudden and strong, so the rabbit may not be able to get back to its litter tray in time. Any rabbit presenting for a breakdown in litter training should be assessed for signs of pain or urinary tract disease.

Fearful

An aversive event or series of events may dissuade the rabbit from using the litter tray (loud noises, unexpected unpleasant stimuli). The rabbit will usually start to urinate consistently in a different location. This will require retraining the toileting behaviour (Advice sheet 10) and preventing the rabbit's exposure to the aversive stimulus.

Territorial

The rabbit's latrine is a key part of its environment. Overzealous cleaning of the litter trays can reduce the scent that the rabbit associates with toileting. Owners should be advised to put a small amount of soiled litter in the cleaned tray, and should not use scented cleaners. If a rabbit urinates outside of the tray and the area is not cleaned appropriately, the rabbit may continue to urinate in the same place (the presence of urine is a strong motivator for the rabbit to continue to eliminate in that site).

Frustrated

If there are insufficient litter trays, or the litter trays are difficult to get to (for example, if only one is provided at a distance from where the rabbit chooses to spend its time), the rabbit may start to use an alternative latrine. Rabbits prefer certain litter tray substrates: if the substrate is changed, the normal association may be lost. Similarly, if the litter trays are heavily soiled, the rabbit may be unwilling to use them. Rabbits frequently eat while defecating, so providing fresh hay in the litter trays encourages the rabbits to spend longer in the tray, so increases the likelihood that they will deposit faecal pellets here.

Seeking social interaction

Rabbits may learn that toileting in an inconvenient place for the owner results in a desirable response from the owner (or at least an interesting one), and may use this as a way to attract the owner's attention. This is more commonly associated with rabbits kept in social isolation, as they are more likely to experience loneliness or boredom, and are more reliant on the owner for social interaction.

Case study 13: Popcorn

Popcorn was a two-year-old neutered male rabbit, whom the owners had had since he was eight weeks old. The owners had never had to formally litter train him – he had reliably toileted in the litter tray since they had brought him home. When he was 18 months old, the owners had decided to bond him to another rabbit, Pumpkin. The bonding process had been uneventful. The owners started to find urine puddles around the house, and initially blamed the new rabbit, Pumpkin, until they observed Popcorn urinating outside the litter tray.

They were very distressed by this, as Popcorn had never shown this behaviour

before. It wasn't clear which aspect of the disruption to Popcorn's routine had caused the behaviour, so the owners were advised to go back to basics and retrain the toileting behaviour as described in Advice sheet 10.

This was successful: the number of 'accidents' reduced dramatically. The owners did find that Popcorn was less likely to use the litter tray if he was in another room with Pumpkin, so they provided a litter tray in each room and confined both rabbits in one room overnight.

To solve a short-term problem, the owner can either place a litter tray in the spot that the rabbit is eliminating, clean the surrounding area and then gradually move the litter tray to a place that is more convenient to them. Or they can restrict access to the area in which the rabbit is urinating and provide a better alternative: an appealing litter tray close by. This may be effective if the rabbit is otherwise well litter trained: if this isn't effective, then the owner should retrain the behaviour in a small area (Advice sheet 10).

Urine spraying

The other form of unwanted urine deposition is urine spraying. Urine spraying is a method of marking territory that is more commonly seen in unneutered male rabbits. Rabbits may spray objects (as part of territorial marking) or their owners or companion (a sexual behaviour).

There is a wide range in the frequency with which unneutered males spray urine, so many owners will not perceive this as a problem. House rabbits that spray urine are more likely to inconvenience their owners than are outdoor rabbits. If the owner is concerned about this behaviour, there are two approaches to dealing with it: neuter the rabbit or manage the problem.

Neutering the rabbit is likely to greatly reduce or eliminate this behaviour. As the behaviour is hormonally driven, it will reduce gradually over four to six weeks after neutering. Some rabbits will continue to spray, and this is more likely if the owner keeps multiple rabbits in different enclosures: the territorial drive here is very high. Additionally, some neutered rabbits may spray when introduced to a new territory, or if their owner has been handling unfamiliar unneutered rabbits, but this behaviour is likely to decline fairly rapidly after neutering.

If the owner wants to keep the rabbit unneutered, or if the rabbit is used for breeding, the owner can reduce the damage caused in several ways. Litter boxes should have high sides so urine isn't sprayed outside the box. Rabbits tend to mark corners and walls, so using plastic wall protectors along walls close to the litter box will allow easier cleaning. Urine marks should be cleaned using an enzymatic, or 'biological', household cleaning solution to reduce territorial marking that is stimulated by the presence of urine. Vinegar may also be effective, as it reacts with some components of the alkaline rabbit urine and may denature some of the volatile compounds. Some rabbits that frequently spray can end up heavily soiled by urine. This is more common in rabbits that are restricted to small environments.

Caecotrophs

Most rabbit owners will know that the rabbit's gut produces two types of pellet: caecotrophs and faecal pellets. If not, you need to work out whether a rabbit that is 'soiling' around the house is depositing faecal pellets or caecotrophs. Unlike faecal pellets, which consist of indigestible material, caecotrophs consist of digestible material. Caecotrophs are eaten directly from the anus (coprophagia), and the nutrients are absorbed on the second pass through the rabbit's gut. Caecotrophs are usually smaller, soft, homogeneous, often covered with mucus and often stuck together like a bunch of grapes. If smeared, the smell is pungent and characteristic. Faecal pellets are usually dry and firm, and don't smell as strongly. Caecotrophs are produced when the rabbit is calm and relaxed, and the rabbit should ingest the caecotrophs directly from the anus. In some instances, the rabbit will pick caecotrophs off the floor to eat, and occasionally will eat the hard, dry faecal pellets: the motivation for this is not clear, but the behaviour is normal.

If the owner reports that a rabbit is leaving caecotrophs outside the litter tray, this may indicate a problem with the rabbit's health. It is unusual for the rabbit to leave caecotrophs: if these are seen, the rabbit may be in pain (so it is uncomfortable for the rabbit to reach its anus), may have underlying disease (altering its motivations) or, intermittently, may have been disturbed or scared when producing the caecotrophs. If caecotrophs are seen on the floor, the rabbit may well have perineal soiling with further caecotrophs, which increases the risk of skin damage and fly strike.

A full clinical examination is essential in this case. Failure to consume caecotrophs is very unlikely to have a behavioural origin. Transitioning the rabbit on to a diet consisting almost entirely of hay or grass, with some vegetables, is likely to increase the palatability of the caecotrophs and reduce this behaviour (Advice sheet 4).

Faeces

Litter-trained rabbits should urinate in the litter tray, but will probably deposit some dry faecal pellets outside the tray to mark their territory. This behaviour is less hormonally driven than urine spraying, so neutering may not always reduce it.

Owners can increase the likelihood of the rabbit defecating in the litter tray by providing plenty of fresh hay and grass in it. Rabbits frequently eat while defecating, so providing food incentivises them to stay in the tray. Owners should also be advised on how to manage this normal behaviour. Confining the unsupervised rabbits to a room where the pellets are easy to clean up will reduce the inconvenience to the owners.

If the rabbit is leaving significant piles of pellets in one specific place (and may be urinating there as well), this indicates a failure of litter training, and should be addressed as described in Advice sheet 10.

Destruction

Destructive behaviour can be motivated by frustration, exploration or inadequate diet, and the rabbit may damage the object by using its paws (digging) or its teeth (chewing or biting). Needless to say, this behaviour can be very distressing to the owner, especially when the items damaged have a high value to the owner. To achieve good resolution of this situation, the cause of the destructive behaviour should be identified. What type of object is being targeted? When is it being targeted? What is the appearance of the damage?

> *The concept of destruction is a human one. The behaviour that is seen as destructive is not a deliberate attack on the owners, as some owners will see it.*
>
> *'When I shut him in his hutch, he has a tantrum.'*
> *'He only eats my iPhone cable because he knows it winds me up.'*

A rabbit usually destroys things because it lacks opportunities to express normal behaviour, but may also be performed by a rabbit that learns that this behaviour elicits a stimulating response from their owner. Rabbits that perform this behaviour for the latter reason typically lack companionship, so try to solicit some form of social interaction in an inappropriate way.

Rabbits showing destructive behaviours respond better to an approach focused on changing the underlying motivation rather than attempting to reinforce an alternative behaviour (or punish the destructive behaviour). This is because the rabbit's motivation to perform these instinctive behaviours is usually very high. Behaviours such as grazing (unwanted when the 'grass' is carpet), removing plant stalks from access routes (unwanted when the 'stalks' are cables) and tree bark stripping (unwanted when the 'bark' is wallpaper) have been strongly selected for during the rabbit's evolution. A rabbit that could not dig a warren in which to live would be at a significant disadvantage, but this behaviour is most upsetting to owners when focused in the corner of the dining room. When these behaviours are focused on an inappropriate object, it shows that the rabbit's environment is not fully meeting its needs.

Some owners will make a loud noise to distract the rabbit when it is performing an unwanted behaviour. This is often sufficient to distract the rabbit in the short term, but is not a long-term solution because the rabbit is likely to perform these behaviours when unobserved. A better short-term distraction is for the owner to call the rabbit over to them – the resultant reward is then in a different location from the object being destroyed.

However, to prevent repetition of these behaviours, the environment needs to be improved.

If the rabbit is kept alone, you should strongly advise the owners to get a companion. Destructive behaviour is less common and usually of a lower magnitude with two rabbits because interactions with the other rabbit use a lot of the animal's time budget, leaving less time for exploratory behaviours. No amount of environmental enrichment can compensate for the lack of a companion, so providing one will reduce destructive behaviour that is motivated by frustration.

Case study 14: He-Man

He-Man was a two-year-old Cinnamon male rabbit that the owners had 'adopted' from a pet shop. The owners had wanted to keep him as a house rabbit, and had purchased a fairly small cage with the idea that He-Man could spend some of his time outside the cage to exercise. However, they were distressed by He-Man's behaviour: he chewed shoes, clothes, carpets, and books. They tried to dissuade this behaviour by throwing cushions at him when he chewed objects, which stopped him in the short term, but he learned to only perform this behaviour when the owner was not in the room.

The owners could not be persuaded to get a second rabbit. There were various aspects of He-Man's welfare that were suboptimal. When the owner understood that He-Man's destructive behaviour was unlikely to improve in his current environment (inadequate housing, infrequent insufficient exercise time, no access to the outdoor environment to express normal behaviours), they decided to get a large hutch and run complex outside.

He-Man moved into an environment with much more space and continual access to a run. He could graze grass

and the owners provided him with tree branches and space to dig. The owners were advised that acquiring a companion for He-Man would further improve his quality of life.

If the rabbit already has a companion, then the next consideration is to optimise the diet. Ensuring that the major component of the rabbit's diet is grass or long-chop hay will help to reduce the motivation to 'graze' on inappropriate substrates. Grass is more palatable to most rabbits than hay, so advising that the owners pick grass or provide easy access to a run on grass will help to reduce the motivation of the rabbit to eat the carpet by providing a more appropriate outlet for this behaviour. Where possible, provide fresh sticks or branches from edible trees (such as fruit trees or willow trees, Figure 5.2) – this allows bark stripping in a way

Figure 5.2: Good pet shops may sell fresh fruit tree branches – these are more likely to be palatable than dried sticks.

Figure 5.3: If rabbits have a permanent enclosure, a concrete box filled with soil can provide a good area in which to dig. Owners may have to change the soil if the rabbits use it as a latrine.

that is more rewarding for the rabbit and more acceptable for the owner.

Digging, obviously, is a behaviour that has been strongly selected in the rabbit's evolutionary history. In the wild, rabbits dig extensive warrens. There is a difference between the sexes in the extent to which they are motivated to dig. Females typically construct large burrows, whereas males are more likely to make scrapes for marking (often urinating and defecating in the dug earth) and in which they lie in hot weather. This is a normal behaviour that should be allowed.

For outside rabbits, owners should be encouraged to provide an area where the rabbits can dig. Recently turned, crumbly soil (adding some sand to a clay soil will lighten it) often stimulates this behaviour. Scattering or even part-burying small pieces of vegetable will also encourage the rabbits to dig in the designated area.

For outside rabbits kept in runs that have wire floors, and for house rabbits, this approach may not be practical. Digging boxes can be set up in high-sided litter trays, or even concrete areas in a permanent enclosure (Figure 5.3). The substrates that seem to be preferred by rabbits are soil and sand. This mixture may be acceptable outside but is unlikely to be practical inside. Some owners use commercially available substrates used in litter trays for cats, but these may cause respiratory problems in rabbits, Large pellets may be uncomfortable for rabbit paws, so may not stimulate digging behaviour to the same extent. Another option is a deep box filled with hay, especially if the owner sprinkles a few pellets through the hay (Figure 5.4).

Figure 5.4: Rabbits may dig in a box filled with hay.

The previous paragraphs focus on making it easier and more enjoyable for the rabbit to express the behaviours in an appropriate situation. The other component of a good behaviour modification plan is to make it harder for the rabbit to show the undesirable behaviour. Certain dimensions of cable (usually fairly small) are much more likely to be chewed through. Taping these against walls or furniture legs makes them less noticeable to the rabbit. Reinforcing wires with insulating tape increases their cross-sectional area, making them harder to bite through.

Finally, the owner needs to understand that when rabbits are kept in an environment that they are not evolved to live in, this can be very stressful for them. Rather than trying to stop them showing their normal behaviours, the owners can help them (and themselves) to cope with the unfamiliar situation by enabling them to express their normal behaviours in ways that are acceptable to their owners.

Self-directed behaviours

Undergrooming

Rabbits should groom themselves frequently. This behaviour is rewarding because it is important for coat and skin health. If a rabbit does not groom itself regularly, its fur and skin will become soiled. This weakens the skin barrier and predisposes the rabbit to dermatitis and fly strike. There are various reasons why rabbits may not adequately groom themselves: long-haired rabbits often struggle to keep themselves clean; painful rabbits may not be able to reach around to groom themselves; and stress may reduce grooming behaviours. Anecdotally, some rabbits develop 'sticky saliva' at times of high stress, which causes tufting of their fur.

When an owner says that their rabbit is not grooming itself, assess the coat condition. Is the coat smooth or rough? Is there any dandruff? Is the coat soiling over the whole coat, or just in one location? What is the condition of the

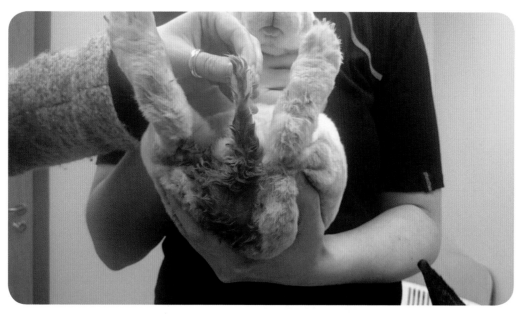

Figure 5.5: Urine scalding can have a variety of causes, both medical and behavioural. Medical causes should always be ruled out first.

paw skin? Is the staining mostly from urine or is the fur matted with caecotrophs? Does the owner ever see the rabbit grooming? Has this behaviour changed at all? Can the rabbit move normally? Can it stand on its back legs?

There are various medical differentials to rule out, including ectoparasites such as fleas and some mites (most rabbits have fur mites, which don't cause signs), dermatitis, urinary tract disease, gastrointestinal tract disease, dental disease, pain or injury (Figure 5.5). Pain is the most common health problem that causes an acute change, and this needs to be ruled out before any behavioural work-up is performed.

As with any behavioural consultation, assess the welfare of the rabbit. Also try to ascertain if there are other abnormal behaviour patterns.

Case study 15: Snowy

Snowy was a three-year-old neutered male New Zealand White, who lived amicably with his brother in a large outdoor enclosure comprising a converted shed and attached run. While his brother was active and healthy, Snowy did not move around much, and spent long periods of time sitting in his litter tray. He did not groom himself well, was often soaked in urine and had pododermatitis on both hind paws. The owners were bathing him when he was soiled (which was very apparent because of his coat colour). A full veterinary work-up had not shown up any abnormalities that might have accounted for this behaviour.

The owners were advised to make Snowy's life as close as possible to that of a wild rabbit, in order to reduce stress and stimulate normal behaviours. Concentrate food was phased out to encourage him to be more active and graze and forage for food. The owner was advised to supply water from a bowl rather than a bottle – eating grass would also increase his water intake to ensure that low-level urinary tract disease was not a contributing factor. In order to increase his motivation to rest outside the litter trays, the owner was advised to provide various areas of deep hay in Snowy's enclosure, and to provide shallow hay in the litter trays. The litter trays were cleaned daily to ensure that he wasn't sitting in wet bedding. Although his body condition score was within the normal range, his owners were advised that reducing his weight to the lean end of normal would reduce the static pressure through his hocks that might improve the pododermatitis.

The urine scalding gradually improved over the next month. He still had occasional days where the owners had to bathe him, but he became more active when he had to find his own food. The owners were pleased with the improvement.

The prognosis for resolving this behaviour is guarded. Where there are obvious welfare problems (rabbit kept in isolation, inappropriate environment or inappropriate diet), or treatable health problems, these should be resolved. The owner should be advised to provide more stimulation and a more normal environment for the rabbit to reduce its inactivity. However, once this normal behaviour is lost, it can be extremely hard for the rabbit to re-establish it.

Owners can try encouraging the rabbit to stand in a shallow puddle for food rewards: the sensation of water on the paws often stimulates the initial paw-flicking behaviour that precedes a groom. The owner can also try gently spraying the rabbit with water on its flanks.

If the rabbit does not start to groom itself, the owner will need to find ways of cleaning it that are not too unpleasant. The rabbit's underside may need to be kept clipped to avoid matting and reduce discomfort, but this procedure itself is aversive. If the lack of grooming leads to persistent health problems, or is a result of pain or disease that cannot be controlled, then euthanasia should be seriously considered.

Overgrooming

In some situations, the rabbit may groom itself excessively. This may be generalised, but is more commonly seen focused on specific locations on the body, resulting in hairless patches or even trauma to the skin. It is relatively uncommon: most excessive grooming is directed at a companion rabbit rather than the individual (see sections on conspecific overgrooming and barbering). In the history, therefore, it is important to ascertain if the damage is caused by the rabbit grooming itself or by being groomed by a companion.

If a rabbit is presented with 'over-grooming', it is vital to rule out a medical cause.

- Parasite infestations or skin disease can cause itching.
- Pain in structures beneath the skin can cause rubbing of the area.
- Continuous exposure to liquid can macerate the skin and cause hair loss – for example, with urine, or tear fluid in lacrimal duct obstruction.
- Intestinal and dental disease have been linked with increased likelihood of overgrooming.

Any rabbit with hairless patches should be referred to a vet to rule out medical causes of this sign.

Grooming is one of the normal rabbit behaviours, and they perform this behaviour (self-care) in balance with other vital behaviours (grazing, sleeping and social interactions). When the rabbit's ability to perform one of these behaviours is limited, its time budget for other behaviours increases. Rabbits that are deprived of access to a companion, access to an appropriate environment or access to a normal diet will spend more time grooming. This behaviour almost always reflects poor welfare, and interventions should focus on improving the rabbit's husbandry.

Overgrooming may cause several problems. The rabbit may remove its fur, cause injury to the skin or even ingest sufficient fur to cause intestinal obstruction (in combination with reduced gut motility).

Unless the skin trauma is bad enough to require treatment, rabbits should not be prevented from grooming. Using Elizabethan collars on rabbits causes substantial distress and prevents them from consuming caecotrophs from the anus. Some owners make a jacket for the rabbit out of socks or tights, which covers the overgroomed area. However, stopping the rabbit grooming itself is damaging to its welfare. If rabbits lack much stimulation, they groom themselves more. Therefore, owners should try to decrease the rabbit's motivation to groom itself by increasing the stimulation from the environment.

If the rabbit needs topical treatment, then the skin may need to be protected for a few days to prevent the rabbit ingesting the medication. In these severe cases, a coat is better tolerated than a collar, but this should be a last resort.

Diet is also an important consideration here – rabbits deprived of access to long forage (hay or grass) are more likely to try to ingest inappropriate substrates that resemble long forage (such as fur). Rabbits should be transitioned on to a diet that consists almost entirely of hay and grass: this will increase the time that the rabbit spends eating (leaving less time for grooming) and will encourage better gut motility (reducing the risk of gastrointestinal problems, even if the rabbit continues to overgroom itself).

Self-mutilation

Self-mutilation is very rare and has mostly been reported in laboratory rabbits (Fox, N., and Bourne, D.). It refers to any self-damaging behaviour, but is usually seen on the pads and digits of the front feet. It may be caused by boredom or frustration, or may have a medical cause (neuropathic pain, trauma, disease). The medical causes should be ruled out before a behavioural diagnosis is made.

If there is no obvious medical cause for the behaviour, then the rabbit's husbandry and welfare must assessed. As for most other behavioural problems, the environment should be improved, the rabbit should be moved on to a grass- or hay-based diet and it should be bonded with a companion (see section on Comparing the effect of diet, environment, and companionship on behaviour at p. 120).

Self-mutilation causes pain and injury to the rabbit. If the behaviour cannot be resolved and the injury is severe, euthanasia should be seriously considered.

Stereotypies

In the 2017 PAW report, 44% of owners reported that their rabbit showed at least one behaviour they wanted to change – 12% of rabbits were reported to bite the bars of their run or hutch repeatedly.

A stereotypy, or stereotypic behaviour, is one that is performed repeatedly in a fixed manner, does not appear to be associated with any particular stimulus and does not have a discernible goal. Lions pacing round and round their zoo

Figure 5.6: Stereotypies focused on the enclosure walls (pawing, digging, reaching up) can indicate insufficiencies in the rabbit's environment.

enclosure are showing a stereotypic behaviour. Stereotypies indicate that the animal is not coping with the stress of its environment – they are much more common in non-domesticated or weakly domesticated species than in species that have been domesticated for a long time. One study reported that 11 per cent of laboratory rabbits showed stereotypic behaviours. Stereotypies include self-mutilation and overgrooming (as previously described). Biting at the bars of the cage, chewing water dispensers, licking walls and floor, swaying the head backwards and forwards, circling the enclosure and pawing at the cage walls (Figure 5.6) have also been described.

In pet rabbits, a study by Normando and Gelli (2011) found that 28.3 per cent of rabbit owners reported stereotypic behaviour problems in their pets, and the prevalence of these behaviours was higher in pets housed more restrictively (pets that were allowed to roam for less than seven hours a day were much more likely to show stereotypic pacing and stereotypic gnawing). The 2017 PDSA report found that 12 per cent of rabbits were 'biting the bars of their run or hutch repeatedly', and that this was a behaviour the owners did not like.

Although stereotypies indicate poor welfare, these behaviours are often non-damaging. They provide a way (although an abnormal way) for the rabbit to cope with its environment. Owners should not try to prevent the rabbit performing the behaviour, as this will further worsen its welfare. Stereotypic behaviour should only be prevented if it is causing physical injury to the rabbit, in which case the behaviour should be limited only as much as is necessary to prevent injury. The owner should provide plenty of opportunity for the rabbit to perform alternative normal behaviours.

Treatment of stereotypies should focus on improving the animal's welfare through its husbandry. If the rabbit is kept on its own, the owner should bond it to another rabbit. If the rabbits are confined for long periods each day, the owner should set the environment up to give them unrestricted access to a sufficiently large area. If the diet is inappropriate, the rabbits should be transitioned on to a better one.

Stereotypies may become habitual, and so may persist even after the welfare of the rabbit has been improved. It can be difficult to determine whether the stereotypy persists because the welfare is still poor (the owner's idea of an 'improved environment' may still not meet the rabbit's needs) or because the behaviour has become ingrained (it has become a habitual 'default' response). In most situations, improving the welfare, increasing opportunities for the rabbit to show normal behaviour and decreasing the time budget that it can spend on stereotypic behaviour should reduce the incidence of the stereotypic behaviour. However, the prognosis for full resolution of the stereotypy should always be guarded.

Food neophobia

A surprising number of owners report that their rabbits show signs of food neophobia (unwillingness to try new foods) when they try to transition them on to a more appropriate diet. In the wild, rabbits taste many different plant species to ascertain whether or not they are likely to be poisonous, and they have very efficient livers to detoxify these small quantities of ingested toxins. This, and the fact that humans rarely use poison to control wild rabbit populations, means that there has not been strong selection pressure for unwillingness to try novel foodstuffs.

In almost every case, the behaviour that owners describe as 'neophobia' is actually satiety – the rabbits are not hungry. A diet high in concentrate food (which is very palatable) will motivate rabbits to eat until they are very full. This decreases their motivation to try novel foods. Advice sheet 4 describes how to change a rabbit's diet. Briefly: the quantity of palatable, fresh grass or hay should be gradually introduced and the quantity of concentrate food should be gradually reduced so the rabbits' hunger stimulates them to start eating a more appropriate diet. There are various suggestions for easing this transition period in the advice sheet.

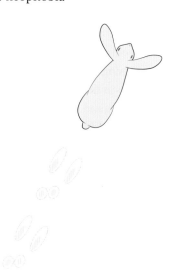

6

Conclusion

With a little bit of knowledge of what rabbits need, and the patience to understand what rabbits want, anyone can resolve unwanted behaviour problems in rabbits.

Clinical animal behaviour is often seen as the preserve of qualified behaviourists, but most of the principles are simple and the techniques straightforward. Having read this book, you are equipped with the tools to make a difference to the lives of pet rabbits.

If you are a rabbit owner who's encountered an unwanted behaviour, consider the different aspects of your pet's welfare. What can you change in its environment to improve its experience? Is that going to be enough to change its behaviour? How are you going to change how you interact with it? How can you make sure that every interaction you have is enjoyable for both you and your rabbit? When you're planning how you are going to change your rabbit's behaviour, it's often useful to write it down: behaviour change happens over days to weeks. Recording a plan helps you to maintain the changes that you make.

If you are a professional who is consulted about unwanted behaviours in rabbits, feel confident to at least open a discussion about it. There is rarely only one way to solve a problem, and you don't need to come to a solution immediately. Initially, ask questions and encourage the owner to talk about the behaviour and the way he or she keeps the rabbit. Think about the Five Domains framework to help you consider the different aspects of the rabbit's welfare. If you haven't got enough information from the history and clinical examination, ask the owner to keep a behavioural diary, or to send you videos of the unwanted behaviour – this will give you both more information, and more time to think. As you collect information, you should start to see aspects of the welfare or the owner–rabbit interactions that could be improved. And when you start to see these contributing factors, explain these to the owner and ask how they could improve the situation for the rabbit and resolve the unwanted behaviour.

You'll have seen that the treatment plans for most unwanted behaviours have the same themes. If you start by improving the rabbit's diet, environment and social contact, you will be most of the way to solving the problem! Having read this book, think about how you are going to improve the behaviour and welfare of the rabbits under your care.

Key points

> **To successfully solve behavioural problems in rabbits, using a clinical approach:**
>
> **1. Repair the husbandry**
> Solving behavioural problems is much easier when the rabbits' husbandry is ideal. This means:
>
> 🐇 **Companionship**
> Rabbits must not be kept without a rabbit companion.
>
> 🐇 **Grass**
> A suitable rabbit diet is mostly, or entirely, grass.
>
> 🐇 **Outdoors**
> Rabbits do better with access to the outdoors.
>
> 🐇 **Grounded**
> Being picked up scares rabbits and worsens behaviour.
>
> 🐇 **Choice**
> It is vital that rabbits have choice over their interactions with humans.
>
> **2. Get in training**
> When rabbits can cope with their environment, they will behave more normally. Training can help them cope.
>
> **3. Set appropriate expectations**
> Rabbits are a relatively recently domesticated prey species. Their motivations, interactions and normal behaviours are very different from cats and dogs. Owners must be aware of this.
>
> **4. Take small steps**
> Set an achievable behaviour modification plan. If many steps are required, build up gradually and highlight results along the way. Remember, the least successful behavioural plan is the one that is not followed ...

Contact details

Please contact me if you'd like to know more about rabbit behaviour, or if you're looking for specific advice on an individual case. I discuss most cases with owners or veterinary surgeons over email (if simple) or over the phone or Skype (if this is easier).

My website is https://bunnybehaviour.wordpress.com and contains links in the dropdown menus to my YouTube videos. The videos show examples of trained behaviours in rabbits and give general advice on clicker training.

My email address is bunnybehaviour@gmail.com – please get in touch!

Acknowledgments

Thank you to the numerous clients who have contacted me through my YouTube channel and website. I have learned so much from your experiences of unwanted rabbit behaviours and your thoughts and feelings during the behaviour modification process. It is heartening to see how many people want to improve their rabbit's behaviour and welfare.

Thanks to Wood Green Animal Shelter, Godmanchester, for letting me take photos of examples of normal rabbit behaviours, novel methods of environmental enrichment and some truly aspirational enclosures!

Finally, huge thanks to Greg, my husband, rabbit photographer and co-author, for his invaluable help with the illustrations and text in this book.

Appendix: Advice sheets

Advice sheet	Intended for
1: Structuring the behavioural consultation	Vets
2: Questions for history taking	Vets
3: Keeping a behaviour diary	Owners
4: Changing the diet of adult rabbits	Owners
5: Improving an outdoor environment	Owners
6: Improving an indoor environment	Owners
7: Constructional approach training	Vets or informed owners
8: Clicker training	Owners
9: Bonding	Owners
10: Litter training	Owners
11: Puzzle feeding	Owners
12: How to interact with your rabbits	Owners
13: Training rabbits to be lifted or moved	Vets or informed owners

Digital copies of these advice sheets for handing out to veterinary clients are available by contacting bunnybehaviour@gmail.com

The following section includes 13 advice sheets on the subjects listed above. Some of these are suitable for giving to owners after a behavioural consult. Some are for vets or more informed owners.

Advice sheet 1: Structuring the behavioural consultation

When you are analysing a problem behaviour, a logical approach will ensure that you collect all necessary information. This also helps the owner to follow your reasoning, making them more likely to agree with the diagnosis and comply with the behavioural modification plan.

- Pre-consultation
  - Ask owner to bring photographs or videos of the problem behaviour, where appropriate.
  - If you perform many behavioural consults, consider using a standard form to collect general details of the rabbit's husbandry.
- Introduction
  - Explain objective for consultation, 'I want to make sure we really get to the bottom of why your rabbit is behaving in this way, and I'd like us to jointly work out how we can best improve the situation for you and your rabbit.'
  - Give a brief overview of plan for the behaviour consultation.
- Detailed history taking
  - General questions.
  - Specific questions.
  - Invite owner to share his or her assessment of why the rabbit is behaving in this way.

- Examination of rabbit and of problem behaviour
  - Clinical examination.
  - Assess photographs or videos, or, if a home visit, assess the environment and the rabbit.
  - Give diagnosis.
- Decide on treatment plan
  - Suggest possible solutions, and invite owner feedback to decide jointly what is achievable.
- Plan appropriate next steps
  - Invite owner to describe how they would judge success. Use this to discuss prognosis.
  - Decide with owner when to next talk to review progress (may be consultation, may be telephone call). Identify what points you would like them to assess in the meantime.
  - Invite owner to contact you if there are problems before the next review meeting.

This structure will help you to organise efficient, effective behavioural consultations. Remember that the least effective behavioural modification plan is the one that is not followed, so make sure that you involve the owner at every step and ensure he or she is invested in the process.

Advice sheet 2: Questions for history taking

A detailed history will help you to make an accurate assessment of the problem behaviour and lead you to a sound diagnosis. There are a lot of questions to ask – don't be daunted! You could consider giving the owners a form containing the general questions before the consultation. This will make your consultation process more effective.

General

1. Signalment
   - How old is your rabbit?
   - What sex is your rabbit? (male or female; neutered or unneutered)
   - What breed is your rabbit?
2. Diet
   - What do you feed your rabbit on? And in what proportions?
   - Roughly how much forage (hay or grass) do you give your rabbit every day? How much does the rabbit eat?
   - Would you describe your rabbit as fussy? Or does it like its food?
   - What treats does your rabbit really like? How often do you give your rabbit treats?
3. Environment
   - Does your rabbit live inside or outside?
   - What bedding do you give your rabbit?
   - How often does your rabbit have access to an exercise area? And for how long?
   - If your rabbit lives inside, to how much of the house does it have access? Does it have a cage in the house? How long is it confined in the cage for?
   - Does your rabbit ever go outside?
   - Can you draw me a diagram of your rabbit's environment, with rough measurements of length/width/height? Please include any other objects that are in the cage.
4. Companionship
   - Does your rabbit live with other companions?
   - Tell me about the companion: age, sex, neutered status, personality.
   - Which rabbit is more likely to lick the other rabbit on the head? Or which rabbit is more likely to be licked?
   - Do your rabbits ever fight?
5. Health
   - Is your rabbit vaccinated?
   - Do you clip your rabbit's claws? Or do you do health checks? How do you catch your rabbits when you do this?
   - Has your rabbit ever been unwell? What did you notice?
   - Is your rabbit currently on any medication?
   - Have you ever taken your rabbit to the vet?
   - Has your rabbit ever had any tooth problems? Or any operations?
   - Does your rabbit have his teeth clipped?

6. Owner interactions
   - How many people live in the house? Any children? How old are the children?
   - Are there any other pets in the house? How does your rabbit interact with those pets?
   - Who interacts most with the rabbit?
   - Who does the rabbit like to interact with most?
   - Would you describe your rabbit as nervous? What is your rabbit scared of?
   - How often is your rabbit picked up off the floor?
   - How does your rabbit tell you when he wants to interact with you?
   - How does your rabbit tell you when he doesn't want to interact with you?
   - Does your rabbit show any defensive behaviours? (lunge, growl, box, bite, etc.)
   - Where on your rabbit's body do you pet him?
   - Do you play with your rabbit? If yes, how do you play with him? And for how long every day?

Specific to the problem behaviour

1. What is the problem behaviour? What worries you about this behaviour?
2. When does it happen? How often does it happen? Is it getting more or less frequent? Where does it happen? With whom does it happen?
3. What triggers this behaviour? What happens immediately before the rabbit shows the behaviour?
4. What do you do when the rabbit shows this behaviour?
5. Does the rabbit behave differently with different people?
6. For how long has the rabbit shown this behaviour pattern? Do you remember when or why it started?
7. How have you tried to change this behaviour before?
8. What would you like to happen in these circumstances?
9. What do you worry will happen if the problem behaviour isn't resolved?

Questions specific to particular problems

- Litter training
  - Does the rabbit use a litter tray? Where is the litter tray located? What sort of litter do you use? How many litter trays do you have?
  - Does your rabbit ever spray urine? Where does he do this?
- Chewing or digging
  - Does the rabbit have access to somewhere that it can dig?
  - Does the rabbit chew its hutch, cage or run?
  - If your rabbit lives in the house, does it chew furniture? Wires? Carpets?
  - Does your rabbit try to dig in areas that you don't want it to dig?

From this, you'll get information on the rabbit's husbandry, the behavioural problem and the owner's thoughts and feelings about the behaviour. This information is invaluable for your diagnostic process.

Advice sheet 3: Keeping a behaviour diary

A behaviour diary is a detailed record of all the times that your rabbit performs an unwanted behaviour over a set period of time. They help to reveal trends and give us information on how to resolve the behaviour. The diary should be filled in by anyone in the house who experiences the unwanted behaviour from the rabbit. Sometimes behaviours are worse with certain people than others, and this is important to know.

Date and time	Where did the rabbit show the behaviour, and to whom?	What happened immediately before the rabbit showed the behaviour?	What behaviour did the rabbit show? (Give as much detail as you can)	What did you do when the rabbit showed this behaviour?	How did the rabbit react?

Advice sheet 4: Changing the diet of adult rabbits

Rabbits are evolved to eat grass and tree shoots. Many behaviour problems occur because the diets provided do not meet their behavioural and physical needs.

Feeding a diet that is entirely grass or hay will help your rabbits in various ways. This is the best thing you can give them for the health of their gut and teeth. Additionally, this helps them to behave in a more normal way – they will spend more time feeding, which satisfies their behavioural needs.

It is very important not to change the rabbits' diet quickly. This plan will transition your rabbits on to a better diet over the next four weeks.

Week 1

Every day, give each rabbit a chunk of hay that is about the same size as him or her. The aim over the next six weeks is that they will eat more and more of it, but they will probably only reduce the amount by about half. You need to give them more than they will eat because there are some parts of the hay that they will almost never want to eat: they need to be able to make their own choice from the hay that you offer.

Reduce the amount of concentrate food that you give to your rabbits by half. Stop feeding this food to them in a bowl; instead, scatter it over the hay or grass so they have to hunt for it. This will increase their interest in the hay and

will prolong the time they spend looking for the food. This is an enjoyable game for them.

Provide small amounts of leafy greens. Weeds such as dandelion or plantain are good, as are leaves from fruit trees like apple or cherry trees. If you can't find these plants, then small quantities of leafy green vegetables from the supermarket are acceptable (brassicas including cabbage, kale, broccoli; spinach; watercress). Do not feed more than a small handful.

Do not feed any of the following: fruits (including apple), root vegetables (including carrots), bread or starchy foods, sweet foods.

Week 2

Continue to feed the same quantity of hay every day. If possible, start to add in grass that you pick for them: initially in small amounts, and then once they start to eat it you can gradually increase the amount of grass that you give. You may find that they eat all the grass that you give them, because grass is tastier than hay (rabbits have evolved to eat grass, not hay).

Reduce the amount of concentrate food that you give them by half again. If your rabbits have been eating muesli foods, then start to feed them half muesli and half extruded nuggets. You will probably find that they are more motivated to eat this food because it is now less available. This is a good thing. This food is not good for their health, and they will be more comfortable and have a better life if they eat more grass or hay.

Continue to give them small amounts of leafy greens: no more than a small handful once a day.

Week 3

Gradually increase the ratio of grass: hay if possible (if you don't have access to grass, then continue with the same quantity of hay). By now, you will find that your rabbits are consuming more hay and grass. If you are worried that your rabbits are losing weight, take them to your vet. Most vets do free weight checks for pets. If your rabbits are not able to eat enough grass or hay, then this may indicate that they already have dental disease.

Start to feed your rabbits a small amount of concentrate food from your hand: about ten small nuggets. If they used to eat muesli, you should now move them entirely on to nuggets instead. This will help them to learn that these foods are 'treats', and it will give you a way to reinforce the behaviours that you want your rabbits to show. Stop feeding them concentrate food in any other situation.

Continue to give them leafy greens: no more than a large handful once a day.

Week 4

Continue to increase the ratio of grass to hay if possible. Only feed your rabbits nuggets from your hand occasionally. You can use these food rewards to reward them for doing the right thing: coming over to you when you call them or going in their cage when you ask them to.

Continue to give them small amounts of leafy greens. If you are feeding weeds or wild plants, then you can increase the amount a bit, but it should always be quite a lot less than the amount of grass or hay that you are feeding: grass or hay is better for your rabbits' teeth.

Week 5 and beyond

Continue as for Week 4.

Congratulations! You have successfully transitioned your rabbits on to a healthier diet! They are very likely to have a longer, healthier, happier life thanks to your actions.

Advice sheet 5: Improving an outdoor environment

In the wild, rabbits live in an area of between 4000 and 20,000 sq m in grassland. They spend a lot of time exploring their environment, grazing on the grass and sleeping underground in their warren. It is not possible to replicate this environment completely for pet rabbits. Therefore, we need to think very carefully about how we can make a smaller space sufficient for them. If the rabbits can perform a wide variety of normal behaviours, the space will seem larger to them. Rabbits need to live in pairs or groups to show many normal behaviours. However, an enriched environment still cannot compensate for a lack of companionship.

When designing an environment for rabbits, you should think about the behaviours that you want the rabbits to be able to express: enclosures can look very different depending on what behaviours they are designed to facilitate. Outdoor environments can allow the rabbits to show important normal behaviours, such as grazing, digging and sunbathing.

Broadly speaking, there are different aspects of the environment that should be considered.

- **Floor space**

  The guideline here is that more space is better. At a minimum, the rabbits should have enough space to run – small dimensions or very tight turns can stop this behaviour. If the floor area of the run is small, consider using mesh tunnels to increase the straight-line distance available for the rabbits to use. Keeping rabbits in tall structures with a small floor area will severely limit their activity, even if the volume of the enclosure is large. The minimum size of enclosure recommended by the RWAF is 1.8 m × 3 m, consisting of a hutch (1.8 m long × 60 cm deep × 60 cm high) permanently connected to a secure run (2.4 m × 1.8 m).

- **Visual barriers**

  Rabbits want to be able to have good visibility of their territory from some aspects, but they also need to have places where they cannot be seen. Consider using tunnels or shelters to create visual barriers, which also serve to make the environment more interesting.

- **Height of the enclosure**

  This is important for two reasons: for the rabbits to be able to reach up and stand on their hind limbs to look around, and for the rabbits to be able to sit on top of objects. In experiments, rabbits work to have access to places to sit that are higher than the surrounding area (even if they don't use them frequently, they like to spend time near them in case of danger). At a minimum, the top of any enclosure should be higher than the height of the tallest rabbit when it stands stretched on its hind legs. Ideally, the enclosure would be taller than this so the rabbit can stretch up even when sitting on top of objects.

- **Areas to dig**

  Digging is normal rabbit behaviour, but one that can obviously lead to escape if not monitored! Many runs have mesh flooring, which prevents the rabbits digging out but also prevents them expressing this normal behaviour. Additionally, it causes unaccustomed pressure on the rabbits' feet (especially if small diameter), so may cause injury. In permanent rabbit enclosures, wire can be sunk into the ground around the perimeter to stop the rabbits digging out, and a skirt of wire can be folded in around the edge as a visual deterrent. The rabbits can then dig freely within the enclosure but cannot dig out of it. If this is not possible, consider providing your rabbits either with a deep soil- or sand-filled litter tray, or allowing them access to an area where they can dig while supervised. Digging is rewarding and provides good physical exercise.

- **Environmental stimulation**
  Rabbits kept outside are exposed to a range of changes throughout the day: changing temperatures, changing light levels, ambient noise, presence of birds, changing weather conditions: all of these provide environmental stimulation. Where possible, having an enclosure that has different microclimates (different levels of shade, shelter, exposure, etc.) will allow the rabbits to regulate their internal state.

- **Toys**
  Different objects are useful to allow different behaviours, and you should consider how you could provide a range of objects that meet different requirements of the rabbit. Many commercial rabbit toys are designed to appeal far more to the owners than rabbits. They are designed to be long lasting (so difficult for rabbits to interact with) and are often not edible (so there is no reward for the interaction). When a rabbit is presented with a new toy, it will usually investigate it to see if it might be

a threat or if it is edible. This behaviour usually lasts for a few minutes at most. If it has no intrinsic reward, the rabbit will then lose interest. Some toys are made from branches, such as willow branches, and are sold as being edible. However, many rabbits don't think that dried branches are particularly tasty. So despite the marketing promise, many rabbits will not engage with these. Rabbits much prefer fresh branches from fruit trees.

Better toys fulfil more than one requirement: they are pleasant to eat, food can be hidden in them and rabbits can move through them or sit on top of them. Examples of such objects are puzzle feeding balls or games, fresh branches from fruit trees, cardboard boxes or plastic tunnels.

The value of objects can be increased in several ways: you can unpredictably hide food in them (so the rabbits have to check them) or you can remove them and replace them so the novelty is maintained. Providing toys that are easily 'reshaped' or damaged (such as cardboard boxes) means that the rabbits can interact more with the object.

- **Shelters**
  Rabbits prefer to have multiple exits from an area where they could be trapped, and frequently will not use a shelter with only one entrance. If you have a permanent shelter in a run, consider cutting an extra entrance. If the rabbits can dig in the enclosure, then you can provide a shelter without a floor and the rabbits can dig other exits. Don't be alarmed if your rabbits do not choose to use a suitable shelter even in bad weather: rabbit fur sheds water very well and many rabbits like to sit outside in the rain.

- **Flooring**
  If the rabbits are outside, the best floor substrate is a lawn: this allows them to

graze and causes a 'normal' amount of pressure and abrasion on their paws and claws. While concrete is easier to clean, it is less yielding and so may contribute to paw injuries. If the floor is concrete, then plenty of hay or straw should be used in many areas to provide areas for the rabbits to rest.

♦ **Feeding**
Ideally, rabbits kept outside should be kept on a lawn, which provides palatable food throughout the year. As a guide, a healthy lawn of 25 sq m should provide sufficient grass to completely support two rabbits year round. Keep an eye on grass levels – if

it becomes very short, consider providing hay. If feeding hay on concrete, change it frequently to prevent mould. And, if you must feed concentrate food, hide it in the bedding, scatter it in the grass or put it in different toys (twisted pieces of newspaper, plastic bottles, cardboard boxes).

Thinking about the design of your rabbits' enclosure will help you to provide opportunities for them to show normal behaviours. This helps them to be happier with each other and with you. The more space and opportunities for normal behaviour that you can provide, the better your rabbits' lives will be!

Advice sheet 6: Improving an indoor environment

In the wild, rabbits live in an area of between 4000 and 20,000 sq m in grassland. They spend a lot of time exploring their environment, grazing on the grass and sleeping underground in their warren. It is not possible to replicate this environment completely for pet rabbits. Therefore, we need to think very carefully about how we can make a smaller space sufficient for them. If the rabbits can perform a wide variety of normal behaviours, the space will seem larger to them. Rabbits need to live in pairs or groups to show many normal behaviours – be aware that an enriched environment still cannot compensate for a lack of companionship.

It is inherently harder to meet some of the rabbits' behavioural needs if they do not have access to the outside. This is because the outside environment has more change (wind, temperature, weather, noises) and is usually closer to their natural habitat (access to large areas of grass for grazing, access to soil to dig in, etc.). However, with some thought you can make a great enclosure for rabbits inside a house.

There are different aspects of the environment that should be considered.

- **Cage**
  Many owners like to provide their rabbits with a cage for litter trays. If your rabbits are ever going to be confined in the cage, then the cage must meet the minimum size of hutch and run complex specified by the RWAF (1.8 m × 3 m, consisting of a hutch [1.8 m long × 60 cm deep × 60 cm high] permanently connected to a secure run [2.4 m × 1.8 m]).

- **Floor space**
  The guideline here is that more space is better. At a minimum, the rabbits should have enough space to run – small dimensions or very tight turns can stop this behaviour. If the room has a small floor space, consider giving the rabbits supervised access to a wider area when you are at home.

- **Visual barriers**
  Rabbits want to be able to have good visibility of their territory from some aspects, but having places where they cannot be seen is also important. Consider using tunnels or shelters to create visual barriers that have another purpose as well.

- **Height of the enclosure**
  If the rabbits have access to a full room, then the height of the enclosure is irrelevant. In the home, if the rabbits are restricted to an area within a room, then it is better to use high barriers to prevent jumping out but to avoid having a roof to the enclosure (providing that you don't need to protect the rabbits from cats and dogs also living in the house). If the enclosure does need a roof, it should be higher than the height of the tallest rabbit when it stands stretched on its hind legs. Ideally, the enclosure would be taller than this so the rabbit can stretch up even when sitting on top of objects.

- **Areas to dig**

  Digging is normal rabbit behaviour, but one that can cause significant damage to flooring and carpets. Consider providing your rabbits either with a deep soil- or sand-filled litter tray, or allowing them access to an area where they can dig while supervised. Digging is rewarding and provides good physical exercise.

- **Environmental stimulation**

  Rabbits kept inside have a much lower level of environmental stimulation than do rabbits kept outdoors (the temperature is usually constant, ambient noise is reduced, there are no changing weather conditions). This means that you need to consider the environment more carefully. Can the rabbits have access to floor-length glass so they can see outside and perhaps sit in the sunlight? (This will not help vitamin D production, as nearly all the ultraviolet light is removed by the glass, but it does enrich their environment).

- **Toys**

  As discussed before, the more Rabbits overheat very easily, so make sure they can also get out of the sunlight and rest in a cooler place. 'stable' environment inside a home results in lower stimulation for rabbits. Toys and objects are therefore more important. Different objects are useful to allow different behaviours, and you should consider how you could provide a range of objects that meet different requirements of the rabbit. Many commercial rabbit toys are designed to appeal far more to the owners than rabbits. They are designed to be long lasting (so difficult for the rabbits to interact with) and are often not edible (so there is no reward for the interaction).

  When a rabbit is presented with a new toy, it will usually investigate it to see if it might be a threat or if it is edible. This behaviour usually lasts for a few minutes at most. If it has no intrinsic reward, the rabbit will then lose interest. Some toys are made from branches, such as willow branches, and are sold as being edible. However, dried branches are significantly less palatable. So despite the marketing promise, many rabbits will not engage with these. Try giving fresh branches from fruit trees instead.

  Better toys fulfil more than one requirement: they are pleasant to eat, food can be hidden in them and the rabbit can move through them or sit on top of them. Examples of such objects are puzzle feeding balls or games, fresh branches from fruit trees, cardboard boxes or plastic tunnels.

  The value of objects can be increased in several ways: you can unpredictably hide food in them (so the rabbits have to check them) or you can remove them and replace them so the novelty is maintained. Providing toys that are easily 'reshaped' or damaged (like cardboard boxes) means that the rabbits can interact more with the object.

- **Shelters**

  Rabbits prefer to have multiple exits from an area where they could be trapped, and frequently will not use a shelter with only one entrance. Cardboard boxes provide good, destructible places to hide or places to sit on, and you can easily cut multiple exits in these. They can also be frequently replaced, which is good for hygiene and provides novelty.

- **Flooring**

  There are a variety of different flooring types inside. Carpet provides good grip, but is more abrasive than grass, and rabbits without access to grass may attempt to graze on the carpet. Vinyl or lino flooring is hard, so can increase pressure on the paws. Indoor rabbits that have weak bones due to lack of exposure to sunlight or a poor diet may be at higher risk of fractures on more slippery flooring, but in rabbits that have good husbandry, this risk is very low. Rabbits adapt relatively quickly and will alter their movement on different floor surfaces. There should be areas with plenty of hay for the rabbits to rest in. Don't only provide hay bedding in litter trays, as this can incentivise the rabbits to sit in damp, soiled hay, which can damage the skin of the paws.

- **Feeding**
  You should make every effort to provide house rabbits with fresh grass: this is more palatable than hay and has the water content for which the rabbit has evolved.

If you choose to feed concentrate food, you can hide it in the bedding, scatter it in the grass, or put it in different toys (twisted pieces of newspaper, plastic bottles, cardboard boxes). House rabbits have a less interesting environment than outdoor rabbits, so encouraging them to spend more time eating hay and grass will reduce boredom.

The considerations above are very important for providing good health and welfare to rabbits housed indoors. However, a lot of this work is unnecessary if the rabbits can spend some of their time outdoors.

Gardens can be made rabbit-secure, and rabbits can be easily trained to use a cat-flap or pet-door. Think carefully about which approach will be right for you and your rabbits.

Advice sheet 7: Constructional approach training

Constructional approach training (CAT for short) is a form of training often used in horses and feral cats. It is a form of negative reinforcement training, where the reward is the withdrawal of the fearful stimulus (which in CAT is the presence of the owner). CAT is useful for rabbits that are very fearful of their owners. When an animal is very stressed, the escape motivation overrides any motivation to find food, so the animal can't be trained with food rewards. In this case, the biggest motivator is to get away from the human, and so this can be used as a reward.

Rabbits that are very fearful of their owners will become more stressed when the owner approaches. They often learn that if they show aggressive behaviour, then the owner will withdraw. In CAT, instead of withdrawing when the rabbit shows aggressive behaviour, the owner withdraws when the rabbit is calm. We don't want the rabbit to get to the stage where it actually shows aggressive behaviour, we need you to learn the signs that your rabbit is getting stressed, and withdraw before it shows the aggressive – these are shown in the diagram.

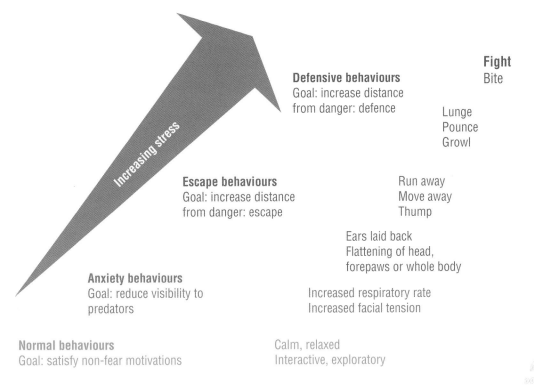

Increasing stress

Fight
Bite

Defensive behaviours
Goal: increase distance from danger: defence

Lunge
Pounce
Growl

Escape behaviours
Goal: increase distance from danger: escape

Run away
Move away
Thump

Ears laid back
Flattening of head, forepaws or whole body

Anxiety behaviours
Goal: reduce visibility to predators

Increased respiratory rate
Increased facial tension

Normal behaviours
Goal: satisfy non-fear motivations

Calm, relaxed
Interactive, exploratory

The goal of constructional approach training is that your rabbit feels confident around you that it neither tries to escape from you nor tries to attack you and that it wants to approach and interact with you. Escape or aggressive behaviours are the end stage of a process where the rabbit has been feeling more and more anxious: when it feels that it has no other option, it shows these behaviours. With this sort of training, we need to see when your rabbit starts to feel anxious, well before it shows the fear behaviours. If you look at the stress escalation ladder diagram above, you can see that the rabbit shows stress behaviours before it is very scared.

Before a rabbit shows the escape or attack behaviours, it shows other, more subtle behaviours. In the wild, the first response of a rabbit to danger is 'freezing', staying very still in the hope that it will not be spotted by a predator. You may see behaviours where the rabbit slightly flattens its body closer to the ground and tenses its muscles (ready to escape if it needs). You may see that the tension in its face increases: its lips look flatter and less rounded and its eyes look bigger and as though they are 'staring'. Its breathing rate may increase. You may also see a change in ear position: if a rabbit has had its ears up, it may bring them closer to its body, or if it was resting when you approached, it may get up slightly and raise its ears. If the rabbit flattens its ears completely to its head, it is very stressed and may attack you: you should never get to this point. You should be aiming to approach to a stage where you see the most subtle changes in the rabbit's behaviour, well below the level of anxiety where the rabbit might show the escape or attack behaviours.

- Initially, you need to find out how close you can get to your rabbit before it starts to feel anxious. You should slowly move your hand towards your rabbit – moving slowly will help you to observe your rabbit's behaviours as they change, and is less threatening to the rabbit.

- Rabbit alters body posture (hunches or tenses), flattens ears and increases tension in the face (behavioural signs of increasing stress, but below threshold for fight-or-flight response).

- Stop moving your hand and stay still (so the 'stress trigger' you represent is no longer increasing in intensity).

- After 10–20 seconds, the rabbit makes a movement, usually lifting the head, turning to look at the trigger or raising the ears (arousal state is reducing, rabbit is relaxing).

- Withdraw your hand (reinforces rabbit for calm behaviour).

- Wait 10–20 seconds before repeating the sequence. You must be sensitive to slight changes in behaviour, and the increases in the intensity of the trigger should be small and incremental.

- You will see that the rabbit starts to show calm behaviours more and more quickly as it learns that this is a way to increase the distance from you. This shows good progress. However, the rabbit's behaviour changes before its emotional state changes: at this point, the rabbit still feels fearful. It is important not to go too fast during this stage, as you can undo the work you've already put in. Continue to watch for signs of anxiety when you're approaching, and be sensitive to calm behaviours so you can withdraw your hand.

- At some point, the rabbit will come forwards to investigate your hand as its confidence builds. At this point, it no longer wants to increase the distance from you – it wants to decrease the distance from you. Keep your hand still and allow the rabbit to investigate your hand on its own terms.

Once the rabbit moves away, you can withdraw your hand.

- When the rabbit has shown interest in coming towards your hand, you should start to positively reinforce this behaviour by holding tasty food rewards in your hand. Holding a piece of leafy green vegetable initially allows the rabbit to approach and eat without needing to interact too closely with your hand. Once it will accept this, hold a small number of pellets in your hand: higher value, but it requires the rabbit to approach more closely.
- Avoid petting the rabbit at any stage during this process: you should 'build' suitable interactions once the rabbit is confident interacting with you (as described in the 'social grooming' section in Advice sheet 12).

With a little effort and a lot of time, you will be well on your way to having happier, healthier rabbits. To reiterate, the goal of constructional approach training is that your rabbit feels confident around you, that it neither tries to escape from you nor tries to attack you, and that it wants to approach and interact with you. There are some excellent YouTube videos of this technique being used in a variety of other species, so take a look at those if you need some more inspiration!

Advice sheet 8: Clicker training

Training your rabbits can help you to manage them without having to pick them up or restrain them. Training is fun for both you and your rabbits, and it can improve their relationship with you. The aim of clicker training is to associate the sound of a click with a good behaviour, and to teach the animal that the click means it will get a reward. This helps you to communicate very specifically with your rabbit about which behaviours you want it to show.

You can see an example of clicker training in action here: https://youtu.be/eYcvf5QrqHk.

Before starting to clicker train your rabbit, it should be comfortable taking small pieces of food (smaller than the nail on your little finger) from your hand. If your rabbit isn't confident with this, start to offer its favourite tasty treats (small pieces of apple or carrot) from the palm of your hand, and then start to offer them from your fingers. Reduce the concentrate food that you give your rabbit in its bowl, and instead start to feed the food from your hand. Your rabbit should start to become more confident in taking food from your hand.

Use a quiet clicker (many of the ones designed for dogs are too loud) or a pop-up jam jar lid to make the 'click' sound.

There are several stages of clicker training.

- **Association of clicker with treat: teaching the rabbit that 'click means food'.**
  - Make a click and then immediately give a food reward. Don't move to offer the food reward until you have made the click.
  - Repeat this sequence 20 times.
  - When the rabbit has learned this association, it will start to expect food when it hears the click: you may see it look at your hands or move forwards.
  - At this point, the rabbit has learned that 'click means food'.
- **Association of clicker with desired behaviour: teaching the rabbit that 'good behaviour causes click'.**
  - Move slightly back from your rabbit, and hold out your hand (without a treat in it). Your rabbit should be interested in you as it is learning that your hand often contains food.
  - When the rabbit moves towards your hand, make a click and then give a food reward from your other hand.
  - Repeat this sequence (holding out your hand and rewarding the rabbit for coming to your hand) until the rabbit is very quick to react to the signal.
  - Then start to move further away from your rabbit before holding out your hand so it has to move further to get a reward.
- **Changing the signal.**
  - A rabbit that approaches to a hand signal is good, but if it comes back to you when you call it, it doesn't have to be able to see you. (Bear in mind that lop-eared rabbits will struggle

to locate a sound because of the ear deformity).

- Whistle your rabbit before you give the hand signal. When your rabbit approaches, click and treat him.
- Repeat this sequence (it can take longer as rabbits are more used to communicating through sight than sound). When you think the rabbit is starting to respond to the whistle, try waiting for longer before giving the hand signal. If your rabbit makes any move in your direction, click and reward this behaviour.
- The first time that your rabbit comes over to you when you whistle, give him a bigger reward (several pieces of concentrate food).
- Once your rabbit has learned one behaviour and is clicker trained, it's much easier to train another behaviour.

Training some simple commands can revolutionise your relationship with your rabbit. You can start to move them around without having to catch them. Rabbits don't like being caught or picked up, so when you stop doing this, you'll find that your rabbits become more confident around you and more interested in interacting with you. Training your rabbits is fun for them and interesting for you!

Advice sheet 9: Bonding

The single best thing you can do to improve your rabbit's life is to bond him to a companion rabbit. Rabbits are highly sociable. Rabbits kept in groups live longer, show more normal behaviours and are healthier.

The easiest way to end up with two rabbits at home is to bring two new rabbits home together: either buying two young rabbits together (who are unlikely to show aggressive behaviour even if they do not know each other), or rehoming two bonded adult rabbits (who are unlikely to show aggressive behaviour as they are familiar with each other). However, if you have one rabbit already, then you'll need to introduce another.

The easiest way to bond two rabbits is to rehome a second rabbit from a good rescue centre. Good rescue centres have the facility to bond two rabbits on neutral territory. You will have to take your rabbit to the rescue centre. The rabbits will be introduced in a novel environment, and then will be supervised until the rabbits have bonded (which may take up to a week). If the rescue centre has several suitable rabbits, they may give you an option to pick three rabbits to introduce to your rabbit, and if your rabbit doesn't like one of the rabbits, they will try another. While the rabbits are at the rescue centre, you should clean their environment at home very thoroughly using a cleaner designed to remove pet urine. You will then take the two rabbits home once bonded, and put them together in the same environment.

If you want to introduce a new rabbit to your current rabbit, there are two methods to do this: fast bonding or slow bonding. Be aware that you are less likely to achieve a successful bonding between the rabbits if you bond them at home than if you bond them at a rescue centre, and it will probably take longer. However, the benefits to the rabbits will be very substantial and hopefully lifelong!

Fast-bonding approach

Before you collect your new rabbit, you need to prepare a suitable unfamiliar territory on which to introduce them. There are two approaches to this. You can either use a completely novel space (such as a shed or a garage), or you can try to clean and rearrange the rabbit's current environment to reduce its territorial drive. The former is much more likely to be successful: rabbits are not daft, and many will show territorial behaviour no matter how much you try to change their environment!

If you are introducing them in an unfamiliar space, set up the environment with fresh hay, clean any litter trays or water bowls very thoroughly and avoid putting in any 'resource focuses' that the rabbit may defend (raised areas, hutches, or food bowls). Provide distractions (scatter tasty food over the floor) and create different spaces for the rabbits to escape to if needed (put various cardboard boxes in to break up the direct sight lines).

If you are introducing them in a space that is familiar to the original rabbit, you need to clean it very thoroughly: use disinfectant and a cleaner designed to remove pet urine. If there are any wooden objects, remove these, as they hold scent very strongly. Change the appearance of the cage to increase uncertainty about territory: move large objects in the room, add new toys or partially cover different parts of the enclosure. Remove resource focuses and provide distractions as described above.

Put both of the rabbits in their new environment. Monitor them continuously for several hours and separate them if you start to see reciprocated aggression.

In the section on 'How to provide companionship', there is a list of the interactions that you may see between the rabbits. If you see aggressive behaviours (described in that section), you should separate them and then move to the slow-bonding approach described below.

Slow-bonding approach

There are three stages to the slow-bonding approach: creating a shared territory without allowing direct contact, and allowing occasional direct contact on neutral territory. When the rabbits are familiar, the final stage is to allow direct contact on the shared territory.

Creating a shared territory without direct contact

Construct a hutch and run complex next to your rabbit's current accommodation. If they have a whole room, you may be able to divide it in two with mesh – preferably with two pieces of mesh so there is a gap of about 10 cm between the mesh. This means that they can't injure each other if they try to fight through the wire. Make sure that there are areas in both runs where the rabbits can see each other, and other areas where they can't.

When you put the new rabbit into one of the enclosures, cover a lot of the mesh with a blanket so there is only a small area where they can see each other. As they become more relaxed and start performing normal behaviours, you can gradually increase the amount they can see.

Switch the rabbits between the pens fairly often so the scent in each becomes a mixture of both rabbits: don't clean the area before you move them.

If you haven't seen signs of aggressive behaviour, reduce the gap between the enclosures so the rabbits can have nose-to-nose contact.

Allowing direct contact on shared territory

After at least a week without any signs of agonistic behaviour, you should start to see the rabbits sitting close to each other on each side of the mesh. You can then try to put them together in their normal environment.

When preparing to introduce the rabbits, remove any 'resource focuses' from both of the environments (i.e. take away any food bowls, block access to the hutch or cage), provide distractions (scatter tasty food over the floor) and create different spaces for the rabbits to escape to if needed (put various cardboard boxes in to break up the direct sight lines). It is also useful to change the appearance of the cage to increase uncertainty about territory: move large objects in the room, add new toys or partially cover different parts of the enclosure.

Then put both rabbits in the same environment. Monitor them continuously and separate them if you start to see reciprocated aggression.

In the section on 'How to provide companionship' 2, there is a list of the interactions that you may see between the rabbits. If you see aggressive behaviours, you should go back several steps: continue to expose the rabbits to each other in unfamiliar situations and start to put them in the shared environment for very short periods (initially five minutes at a time, then gradually start to increase it).

Most rabbits can be bonded; however, there are some pairs that just don't seem to work. If, after a month, the rabbits are still fighting, you may need to admit defeat. If you think that your rabbit may be difficult to bond, speak to a good rescue centre. They may be able to help you bond two rabbits (if you rehome one from them), and may be able to give your rabbit different pairing options to increase the likelihood of a successful bond. Alternatively, if they don't have success themselves, they may be willing to let you try a slow-bonding process and take back the new rabbit if the process is not successful.

Once the rabbits are bonded, separate them as little as possible. If you need to take one to the vets, take both. If one needs hospitalisation, hospitalise both. A good bond is fairly strong but it can be disrupted, and it is very frustrating to have to go through all of those steps again!

Bonding rabbits can be time-consuming and frustrating. However, it is so important for their happiness. Rabbits are really social animals – seeing them grooming each other, playing and exploring together is very rewarding. Although the bonding process can take several weeks, they will hopefully be together for many years. Good luck!

Advice sheet 10: Litter training

Wild rabbits deposit urine and faeces mostly in an area called a 'latrine'. Pet rabbits also choose to do this. This means that you can train them to use a litter tray as their latrine, and so reduce the number of faecal pellets that you'll find around their living area. A litter-trained rabbit should only urinate in the litter tray, and should leave most of the faecal pellets there. However, most rabbits will still deposit occasional faecal pellets elsewhere – this is because they use them as territorial markers.

When litter training rabbits, you need to provide them with an area that meets their needs as a latrine, and you need to decrease their motivation to toilet elsewhere. Practically, this means that you want to set up a good, appealing litter tray and restrict their access elsewhere until they are using the litter tray. Then the scent markers of the litter tray will increase the likelihood that the rabbits continue to use this site (Magnus 2010).

This process is useful when introducing a new rabbit to the house, and if there is a breakdown in the rabbits' litter-trained behaviour (for example, if they are suddenly scared when using the litter tray and start to toilet elsewhere). If a previously house-trained rabbit starts urinating elsewhere, this can be a sign of illness, so you should see your vet.

- The rabbits should be confined in a small area for about 48 hours, such as an indoor cage. The floor should be covered in hay or straw, with newspaper underneath. The rabbits can then choose a particular area as a latrine: once they are reliably urinating in the same place, then you should put a litter tray in this area. Put the urine-soaked newspaper in the litter tray.

- Rabbits like to eat while they toilet, so fresh hay is a good top layer for the litter tray (this also increases the time that the rabbits spend in the litter tray, which may reduce the faecal pellets that they leave elsewhere). The under layer should be very absorbent, so newspaper or cat litter can be used underneath. Try not to change the litter that you use, as this may discourage your rabbits from using the tray.

- If you have several rabbits, make sure there is enough space for all rabbits to toilet at the same time without competition. The standard advice for litter trays is that you should have one per rabbit plus one extra (so for two rabbits, have three litter trays).

- Leave the rabbits in the small area for another 24 hours after introducing the litter tray, so they become accustomed to using it.

- Gradually increase the area to which the rabbits are allowed. Overnight, it is often easier to confine the rabbits to the same room as the litter trays. Rabbits that sleep a long distance away from their litter trays may start to toilet elsewhere for convenience. If the rabbits routinely have access to several rooms, consider providing a litter tray in each room.

- Clean out the litter tray regularly, but not too often: it needs to be hygienic but over-cleaning the litter tray will reduce the rabbits' motivation to use it as a latrine, as it won't 'smell right'. You can also put a small amount of soiled litter into the tray when it is clean to act as a scent marker.
- If the rabbits frequently urinate over the side of the litter tray, you might need to get one with higher sides. When a rabbit urinates, it sticks its bottom out so it doesn't urinate on its back feet. If the sides of the litter tray are low, the rabbit may have its paws in the litter tray but its bottom over the side.
- If the rabbits start to urinate next to the litter tray, they may dislike something in the litter tray (if you have changed an aspect of it), or they may struggle to get in and out (if they are in pain).
- If the rabbits begin to urinate elsewhere, confine them to a smaller area again to re-establish the habit. Clean any urine marks with an enzymatic cleaner designed for pet urine (bleach is not effective) or with a biological washing powder, and rinse well. Rabbits often urinate where they can smell urine: this is part of their communication behaviour.
- If the rabbits start to urinate on a bed or chair, you should prevent them spending long periods of time in that location. A strong recall is useful to redirect their attention. When one of the rabbits jumps on to that area, call the rabbit as you walk towards its cage and give it a treat by its cage when it comes to you. You are then rewarding the rabbit for spending time near its cage: if it needs to urinate, it is also then in the right place.

Litter training rabbits is relatively straightforward, and will make your life easier! Bear in mind that the rabbits can relapse at times of stress. If this happens, just perform these steps again to help them relearn the right habit.

Advice sheet 11: Puzzle feeding

Wild rabbits spend most of their day grazing. They have a large area (4,000–20,000 sq m) that they explore to find new, tasty plants. Pet rabbits do not have access to this amount of space, so can become bored and inactive. You can help to make their enclosure more interesting by hiding some of the food that you provide so they have to find it. This food might be pelleted food (though remember that most rabbits are healthier with little or no pelleted food), small pieces of vegetable, or leaves of herbs and greens. This is called puzzle feeding.

Providing food in a bowl is easy for you, but is unnatural for the rabbits. It can contribute to behavioural problems such as aggression around the food bowl (directed at the rabbit's owner or companion rabbit), and may contribute to boredom. You can provide food to your rabbits in a number of ways – many owners give rabbits their whole daily ration of concentrate (at most, an egg cup full) through puzzle feeding. When you give them their green vegetables and herbs, you can also hide them under the hay or in different parts of their enclosure.

- **Commercial puzzle feeders**
  Many of the puzzle feeders available for dogs and cats are also suitable for rabbits. Opaque feeders (where the food cannot be seen) are acceptable to rabbits: they cannot see directly in front of their nose anyway. Rabbits have less ability to manipulate objects with their paws than do cats and dogs, but can move objects with their teeth.

Simple puzzle feeding balls are usually well accepted. Small, extruded nuggets make the best rewards in puzzle feeders as they easily fall through the holes in the feeder.

- **Homemade toys**
  The advantage of many homemade toys is that the rabbits can damage them or tear them apart. This isn't possible with many commercial feeders because they are designed to resist damage. You can hide food in paper- or cardboard-based objects (egg boxes, paper bags, newspaper twisted around the food, kitchen roll tubes), you can cut holes in plastic bottles (so the food comes out when the rabbits nudge the bottle) or you can suspend small parcels of food from string (unpredictable movement and difficult to access).

- **Scattering food in hay**
  Scattering concentrate food or shredded green vegetables into fresh hay has several advantages. The action of finding the food within the hay is a natural one, so very easy for rabbits to perform, but it also increases the rabbits' interest in the hay or grass, so can be a useful way for owners to increase the motivation of their rabbits to eat more forage.

- **Providing fresh tree branches**
  In the wild, rabbits browse as well as graze, so strip bark off trees to eat it. While you can buy dried bark to feed rabbits, it's much less palatable and doesn't provide the behavioural stimulus of fresh branches.

The tastiest branches come from fruit trees (apple, pear, cherry, apricot).

- **Training**

  You can either provide food in an interactive toy, or you can provide food yourself in an interactive way. You can hand feed a rabbit its food, which will make it more interested in you, and then you can start to teach it to do simple behaviours. Try calling it over to you by a whistle or its name, and giving it some food when it gets to you. If you are doing some form of training, rather than just giving food, the rabbits will find you even more rewarding and you will provide more interest in their day.

Puzzle feeding is stimulating for your rabbits and fun for you to watch! Your rabbits will learn to find the food more quickly as they practise. Puzzle feeding is an excellent way of improving your rabbits' physical and mental health.

Advice sheet 12: How to interact with your rabbits

We often expect rabbits to conform to our views on how pet rabbits should behave: we think that they should enjoy being picked up and cuddled, should be happy to be stroked when we stroke them and should play games with us like a dog would. However, rabbits are eaten by birds of prey, foxes and many other species, and being picked up usually means that they are about to be eaten. Therefore, rabbits instinctively don't like to be picked up. They like to have choice over how they interact. They also like to play, but their 'play' behaviours are not very obvious to us.

So how should you interact with our rabbit? There are various things that you can do.

- **Hand feeding your rabbit**
  When you feed a rabbit from your hands, you teach it that your hands can be a source of pleasurable things. Often, rabbits develop negative associations with human hands because hands are used to restrain them or pick them up. If your rabbits show aggressive behaviours towards your hand, calling them over to give them food from your hand is the first step in reducing their fear.

- **Social grooming**
  Rabbits 'ask' each other if they want to be groomed by lowering their heads. You can mimic this behaviour by putting your closed fist gently in front of your rabbit's nose. If the rabbit doesn't lower its head, then you should take your hand away. If the rabbit lowers its head, you can stroke it. Rabbits like to be stroked along the front and top of the head, around the eyes and along the ears. They often tolerate being stroked on the back (this can be a sexual behaviour between rabbits) but prefer being stroked on the head. If the rabbit withdraws, then you should stop stroking it and back off. This teaches the rabbit that you are paying attention to its communication, so it is less likely to need to show aggressive or defensive behaviours.

- **Training your rabbits**
  If you can hand feed your rabbits, you can think about training them to do simple behaviours. There are other advice sheets in this book on clicker training and constructional approach training for ideas on how to go about this.

- **Giving your rabbits novel objects**
  Play behaviours in rabbits are much more common in young rabbits, and usually involve running and chasing. However, investigative behaviour is rewarding for rabbits at all ages. Try giving your rabbits food hidden in toys (see the advice sheet on puzzle feeding) or novel objects for them to explore. This is stimulating for your rabbits and fun for you to watch.

- **Spending time near your rabbits**
  Rabbits are very social animals and like to spend time in proximity to other rabbits. Even if your rabbits are not snuggled up next to you, allowing them in the room

while you are watching television or reading will give them a chance to spend time with you. Many owners will find that their house rabbits follow them round the house so they can rest in the same room as their owners.

When you only interact with your rabbits in ways that they like, you will find that they become more confident around you. They will trust you more. They will see you more like another rabbit than a potential predator. This will really improve your relationship with them.

Advice sheet 13: Training rabbits to be lifted or moved

Rabbits don't enjoy being picked up – they are eaten by birds of prey, foxes and many other species. Being picked up usually means that they are about to be eaten. Rabbits instinctively don't like to be picked up. However, there are occasions when a rabbit needs to be picked up, and so training them to tolerate being picked up can reduce stress at this time. You must understand, however, that training a rabbit to tolerate this behaviour does not mean the rabbit will enjoy it. Many rabbits will submit to being picked up in exchange for a good food reward.

Small rabbits and large rabbits can be picked up in different ways. Here, 'small' rabbits are those that can be scooped up in cupped hands. 'Large' rabbits are those that are too long in the body to be lifted like this.

Rabbits dislike their paws and underside being touched by human hands. This affects how we pick them up – we can lift small rabbits primarily through pressure on their sides, and we can lift larger rabbits in towels. All sizes of rabbits can be trained to go into a carry cage on command and be lifted in that (this also needs to be trained gradually as rabbits dislike feeling trapped or confined.

Training a rabbit to be lifted in a carry cage

- Leave the open cage in the rabbit's environment and place food and bedding inside it. If the cage opens at the top, then rotate the cage so the lid is now on the side so the rabbit can easily get in and out.
- Once the rabbit is comfortably going in and out of the cage when food is provided (most rabbits prefer multiple exits from a confined space, so the rabbit is unlikely to rest in the cage), then start to throw food in to encourage the rabbit to enter the cage.
- Say 'cage' and make a hand gesture into the cage. If your rabbit is used to getting food in the cage, it should go into the cage. Give it food when it does so.
- Repeat this, but start to touch the door of the cage before giving the rabbit food.
- Gently close the door, feed the rabbit through the mesh of the door and then open the door. You may find that the rabbit hops out at this point. That's fine, move away, and then try again in a few minutes.
- Once the rabbit is calm and relaxed when you close the door, try closing the door, moving the cage slightly, giving some food and then opening the cage.
- Once the rabbit is calm and relaxed when the cage is moved, close the door, lift the cage briefly off the floor, give some food and then open the cage.
- As you are unlikely to need to transport rabbits frequently in the cage, the goal of this exercise is to train the rabbit to willingly go into the cage and be calm and relaxed when the cage is moved. Training the rabbit to tolerate longer and longer

periods of being carried in the cage is unlikely to be productive – the rabbit is likely to become frustrated and form negative associations with the cage.

Training a small rabbit to be picked up

Only try this with a clicker-trained rabbit, as rabbits dislike the sensation of being picked up by human hands more than they dislike being confined in a box. Your rabbit should be relaxed and comfortable around you, and should not show any fearful behaviour towards your hands. If it does, you need to stop picking it up at any time and first treat the fear-related behaviour. Your rabbit should also lower its head when it wants to be groomed by you: this will become an important signal so you know if the rabbit is happy to be picked up.

- To pick up the rabbit, you will need to use both hands and put one hand on each side of the rabbit. Use a cue that sets you up to do this: I put both hands in front of the rabbit and open them like a book. A rabbit that is used to being groomed will lower its head. You can then move the hands alongside the rabbit without touching it, click and then give a treat.

- When the rabbit is reliably lowering its head when you give the cue, and staying relaxed with your hands on each side of it, you can increase the intensity of the stimulus. Move your hands to touch the rabbit high up on its flanks, click, then remove your hands and give the rabbit a treat. This is an unfamiliar sensation to your rabbit, and it may initially move away. Give it a few minutes to relax before trying again (always waiting for the rabbit to lower its head before you move your hands). The

pressure should be light, high up on the body and quickly removed.

- When the rabbit is relaxed when you apply pressure to its sides, you can gradually move your hands lower (the goal is to be able to cup the rabbit in your hands without needing to put your hands under its paws) and apply more pressure for longer. Start by increasing the pressure: when the rabbit is relaxed with this, you can apply the pressure for longer. Once the rabbit is relaxed here, then start to move your hands lower. Always give the hand signal and wait for the rabbit to lower its head before you move your hands.

- When the rabbit is relaxed with those interactions, gently lift your hands slightly: taking some of the rabbit's weight but not lifting its paws off the ground. This again is a novel sensation, so you may need to spend a while on this stage, or go back a stage, until the rabbit is calm and relaxed.

- Once the rabbit is calm and relaxed, you can start lifting it completely off the floor and then down again. Gradually progress to lifting it, moving it sideways and putting it back on the floor. Ensure that you give very desirable rewards here to motivate the rabbit to continue to engage.

- As you should set up the environment and train the rabbit to come to you, you shouldn't need to frequently pick up your rabbit. The goal of this exercise is to train the rabbit to 'agree' to being picked up in certain situations in response to an obvious signal. Try to avoid picking it up without giving this cue: this will reduce the rabbit's trust in you. Training a rabbit to be picked up doesn't mean it will enjoy the sensation, so training it to be carried around or picked up frequently may increase its frustration and reduce its willingness to engage with you.

Training a large rabbit to be picked up in a towel

Only try this with a clicker-trained rabbit, as most rabbits don't like the sensation of being lifted. Rabbits don't like the sensation of human hands touching their paws, and with large rabbits, it is not possible to pick the rabbit up with your hands without touching their paws (unlike small rabbits, see above). Therefore the aim of this exercise is to train your rabbit to be comfortable and relaxed when lifted in a towel.

- Select a cushion that is large enough for your rabbit to sit comfortably on and a towel that has a width larger than the length of your rabbit from the tip of the nose to the end of the tail. Place the cushion on the floor and cover it with the towel.

- Initially, use a target or your finger to encourage the rabbit to put its paws on the cushion, facing you. Click and reward this behaviour. As the rabbit becomes more confident, it should start to hop willingly on to the cushion towards you. Continue to click and reward this behaviour.

- When the rabbit is confident hopping on to the cushion, start to lift the edges of the towel before clicking and rewarding the rabbit. Gradually increase the amount that you lift the edges until the rabbit is enclosed in the towel. Click and reward at each stage.

- Now lift the towel slightly at the edge and slightly move the cushion. Click and reward the rabbit for staying still. This helps the rabbit to become accustomed to the feeling of a less-secure foothold (but is still more secure than just lifting the rabbit in the towel). As the rabbit becomes more confident, gradually increase the movement of the cushion.

- Once the rabbit is relaxed when the cushion is moving, go back to just lifting the towel and build up again (clicking and rewarding the rabbit for staying still). Start to take a bit of the rabbit's weight when you lift the towel. Take this stage slowly. The aim is to be able to pick the rabbit up in the towel and move it or put it into a top-opening cage.

- As you should set up the environment and train the rabbit to come to you, you shouldn't need to frequently pick up your rabbit in a towel. Try to avoid picking it up in other situations: this will reduce the rabbit's trust in you. Training a rabbit to be picked up doesn't mean that it will enjoy the sensation, so training it to be carried around or picked up frequently may increase its frustration and reduce its willingness to engage with you.

It can be hard to train this behaviour because rabbits really dislike being picked up. Only try this with rabbits that already know a number of commands (they will understand that they can get a food reward, and will already trust their owners more). Try to avoid picking up your rabbit as much as possible. Train them to go into their enclosure, or come to you on command. However, sometimes you will need to pick them up, and if you've trained them to tolerate this, they will be much less stressed.

Glossary

Affect – the experience of feeling or emotion, part of the process of an animal's interaction with stimuli.

Affiliative behaviour – behaviours that promote group cohesion (friendly/positive gestures), such as mutual grooming or resting in proximity.

Aggressive behaviour – behaviours that cause stress or physical harm to another individual.

Agonistic behaviour – any social behaviour related to fighting. This includes overtly aggressive behaviour, but also includes threats, displays, retreats, placation and conciliation.

Associative learning – learning that occurs when an animal realises that there is a link between two events, or a link between an event and a behaviour. Also called 'conditioning'.

Aversive – causing avoidance of a thing, situation or behaviour by using an unpleasant or punishing stimulus.

Bridging stimulus – a marker that identifies the desired response and 'bridges' the time between the response and the delivery of the primary reinforcer.

Classical conditioning – form of associative learning, where an animal learns that a neutral stimulus is paired with a potent stimulus, and so it has an innate response when the neutral stimulus occurs.

Clicker training – form of positive reinforcement training using a bridging stimulus.

Constructional approach training – form of negative reinforcement training using withdrawal of a person to reinforce relaxed behaviour when approached.

Motivator – something that provides a stimulus for an animal to do something.

Non-associative learning – learning that occurs when an animal is repeatedly exposed to a single stimulus, and that stimulus is not linked with anything else.

Operant conditioning – form of associative learning, where an animal learns to modify its behaviour based on the consequences of that behaviour, and learns 'discriminative stimuli', or 'antecedents', that alter that behaviour.

Punisher – a consequence that decreases the likelihood of a behaviour being repeated.

Reinforcer – a consequence that increases the likelihood of a behaviour being repeated. Often food rewards are used as reinforcers during training.

Response – an observable activity that occurs following a stimulus.

Stimulus – something that causes a response in an animal. A stimulus may be internal or external.

Further reading

Association of Pet Behaviour Counsellors. www.apbc. org.uk/tips/rabbit [accessed 05/03/17]

Bradbury, A.G., 2013. 'Bunny Behaviour'. https://bunny behaviour.wordpress.com [accessed 22/02/17]

Crowell-Davis, S.L., 2007. Behavior problems in pet rabbits. *Journal of Exotic Pet Medicine*, *16*(1), pp. 38–44.

Crowell-Davis, S.L., 2010. Rabbits. In V.V. Tynes (ed.), *Behavior of Exotic Pets* (pp. 69–77). Wiley-Blackwell, Oxford, UK.

Pryor, K., 2002. *Don't Shoot the Dog!* The new art of teaching and training. Revised Edition. Ringpress. Reading, Great Britain.

Rabbit Welfare Association & Fund. www.rabbitwelfare.co.uk [accessed 05/03/17]

Bibliography

Adamec, R.E., 1976. The interaction of hunger and preying in the domestic cat (Felis catus): an adaptive hierarchy?. *Behavioral Biology*, *18*(2), pp. 263–272.

Albonettl, M.E., Dessi-Fulgherl, F. and Farabollinl, F., 1990. Intrafemale agonistic interactions in the domestic rabbit (Oryctolagus cuniculus). *Aggressive Behavior*, *16*, pp. 77–86.

Animal Welfare Act 2006. www.legislation.gov.uk/ukpga/2006/45/contents [accessed 20/02/2017]

Belyaev, D.K., 1979. Destabilizing selection as a factor in domestication. *Journal of Heredity*, *70*(5), pp. 301–308.

Boissy, A., Manteuffel, G., Jensen, M.B., Moe, R.O., Spruijt, B., Keeling, L.J., Winckler, C., Forkman, B., Dimitrov, I., Langbein, J. and Bakken, M., 2007. Assessment of positive emotions in animals to improve their welfare. *Physiology & Behavior*, *92*(3), pp.375–397.

Bonas, S., McNicholas, J., & Collis, G. M. (2000). Pets in the network of family relationships: An empirical study. In A. L. Podbersek, E. S. Paul, & J. A. Serpell (Eds.), *Companion animals and us: Exploring the relationships between people and pets* (pp. 209–236). New York, NY, US: Cambridge University Press.

Bradbury, A.G., 2016. Managing conspecific overgrooming in rabbits. *Veterinary Record*, *178*(12), pp. 298–299.

Bradbury, A.G. and Dickens, G.J.E., 2016. Appropriate handling of pet rabbits: a literature review. *Journal of Small Animal Practice*, *57*(10), pp. 503–509.

Bradbury, A.G. and Dickens, G.J.E., 2016. Should we advocate neutering for all pet rabbits?. *Veterinary Record*, *179*(25), pp. 654–655.

Calvete, C., Estrada, R., Lucientes, J., Osacar, J.J. and Villafuerte, R., 2004. Effects of vaccination against viral haemorrhagic disease and myxomatosis on long-term mortality rates of European wild rabbits. *Veterinary Record*, *155*(13), pp. 388–391.

Cats & Rabbits & More. Cats and rabbits together. www.catsandrabbitsandmore.com/cats___rabbits_together [accessed 15/01/17]

Courcier, E.A., Mellor, D.J., Pendlebury, E., Evans, C. and Yam, P.S., 2012. Preliminary investigation to establish prevalence and risk factors for being overweight in pet rabbits in Great Britain. *Veterinary Record*, *171*(8), pp. 197–197.

Coureaud, G., Charra, R., Datiche, F., Sinding, C., Thomas-Danguin, T., Languille, S., Hars, B. and Schaal, B., 2010. A pheromone to behave, a pheromone to learn: the rabbit mammary pheromone. *Journal of Comparative Physiology A*, *196*(10), pp. 779–790.

Cromer, L.D. and Barlow, M.R., 2013. Factors and convergent validity of the pet attachment and life impact scale (PALS). *Human-Animal Interaction Bulletin*, *1*(2), pp. 34–56.

D'Eath, R.B., Tolkamp, B.J., Kyriazakis, I. and Lawrence, A.B., 2009. 'Freedom from hunger' and preventing obesity: the animal welfare implications of reducing food quantity or quality. *Animal Behaviour*, *77*(2), pp. 275–288.

d'Ovidio, D., Pierantoni, L., Noviello, E. and Pirrone, F., 2016. Sex differences in human-directed social behavior in pet rabbits. *Journal of Veterinary Behavior: Clinical Applications and Research*, *15*, pp. 37–42.

Davies, O., Dykes, L., 2000. Surface attraction: skin problems in rabbits. *Rabbiting On*. www.rabbitwelfare.co.uk/resources/content/info-sheets/sorehocks.htm [accessed 02/01/17]

Dúcs, A., Bilkó, Á. and Altbäcker, V., 2009. Physical contact while handling is not necessary to reduce fearfulness in the rabbit. *Applied Animal Behaviour Science*, *121*(1), pp. 51–54.

Fairham, J. and Harcourt-Brown, F.M., 1999. Preliminary investigation of the vitamin D status of pet rabbits. *Veterinary Record*, *145*(16), pp. 452–454.

Farm Animal Welfare Council, 2009. Five Freedoms. http://webarchive.nationalarchives.gov.uk/20121007104210/www.fawc.org.uk/freedoms.htm [accessed 18/02/2017]

Fox, N., and Bourne, D., 2017. Self-mutilation in rabbits. *Wildpro*. http://wildpro.twycrosszoo.org/S/00dis/PhysicalTraumatic/Self_mutilation_rabbits.htm [accessed 08/02/17]

Fox, N., 2012. Barbering and excessive grooming in rabbits. *Wildpro*. http://wildpro.twycrosszoo.org/ S/00dis/ PhysicalTraumatic/Barbering_in_rabbits. htm [accessed 08/02/17]

German, A.J., 2015. Style over substance: what can parenting styles tell us about ownership styles and obesity in companion animals? *British Journal of Nutrition*, *113*(S1), pp. S72–S77.

González-Mariscal, G., Melo, A.I., Zavala, A. and Beyer, C., 1990. Variations in chin-marking behavior of New Zealand female rabbits throughout the whole reproductive cycle. *Physiology & Behavior*, *48*(2), pp. 361–365.

Hansen, L.T. and Berthelsen, H., 2000. The effect of environmental enrichment on the behaviour of caged rabbits (Oryctolagus cuniculus). *Applied Animal Behaviour Science*, *68*(2), pp. 163–178.

House Rabbit Society, 1997. Overgrooming. http:// rabbit.org/overgrooming [accessed 14/03/17]

Illera, J.C., Silvan, G., Lorenzo, P., Portela, A., Illera, M.J. and Illera, M., 1992. Photoperiod variations of various blood biochemistry constants in the rabbit. *Revista Espanola De Fisiologia*, *48*(1), pp. 7–12.

Lebas, F., Coudert, P., De Rochambeau, H. and Thebault, R.G., 1997. *The rabbit: husbandry, health and production*. (New revised version). FAO Rome.

Magnus, E., 2010. Litter training your rabbit. *Association of Pet Behaviour Counsellors*. www.apbc.org.uk/ articles/rabbitlitter [accessed 05/03/17]

Mancinelli, E., Keeble, E., Richardson, J. and Hedley, J., 2014. Husbandry risk factors associated with hock pododermatitis in UK pet rabbits (Oryctolagus cuniculus). *Veterinary Record*, *174*(17), pp. 429–429.

Masoud, I., Shapiro, F., Kent, R. and Moses, A., 1986. A longitudinal study of the growth of the New Zealand white rabbit: cumulative and biweekly incremental growth rates for body length, body weight, femoral length, and tibial length. *Journal of Orthopaedic Research*, *4*(2), pp. 221–231.

McBride, E.A., 2017. Small prey species' behaviour and welfare: implications for veterinary professionals. *Journal of Small Animal Practice*, *58*(8), pp. 423–436.

McBride, E.A., Day, S., McAdie, T.M., Meredith, A., Barley, J., Hickman, J. and Lawes, L., 2006. Trancing rabbits: relaxed hypnosis or a state of fear? In *Proceedings of the VDWE International Congress on Companion Animal Behaviour and Welfare*, pp. 135–137. Flemish Veterinary Association.

Mellor, D.J., 2016. Updating animal welfare thinking: moving beyond the 'Five Freedoms' towards 'a life worth living'. *Animals*, *6*(3), p.21.

Melo, A.I. and González-Mariscal, G., 2010. Communication by olfactory signals in rabbits: its role in reproduction. *Vitamins & Hormones*, *83*, pp. 351–371.

Meredith, A.L., 2010. The rabbit digestive system: a delicate balance. *Rabbiting On* 7. http://rabbit welfare.co.uk/pdfs/ROWinter10p7.pdf [accessed 01/01/17]

Meredith, A.L., 2013. Viral skin diseases of the rabbit. *Veterinary Clinics of North America: Exotic Animal Practice*, *16*(3), pp. 705–714.

Meredith, A.L., Prebble, J.L. and Shaw, D.J., 2015. Impact of diet on incisor growth and attrition and the development of dental disease in pet rabbits. *Journal of Small Animal Practice*, *56*(6), pp. 377–382.

Mullan, S.M. and Main, D.C., 2007. Behaviour and personality of pet rabbits and their interactions with their owners. *Veterinary Record*, *160*(15), pp. 516–520.

Normando, S. and Gelli, D., 2011. Behavioral complaints and owners' satisfaction in rabbits, mustelids, and rodents kept as pets. *Journal of Veterinary Behavior: Clinical Applications and Research*, *6*(6), pp. 337–342.

Panksepp, J., 2004. *Affective neuroscience: the foundations of human and animal emotions*. Oxford University Press, New York, USA

Paul-Murphy, J., 2006. What is pain? *House Rabbit Society*. www.rabbit.org/health/pain.html [accessed 02/01/17]

PDSA, 2016. Animal wellbeing report. https://www. pdsa.org.uk/~/media/pdsa/files/pdfs/veterinary/ paw-reports/pdsa-paw-report-2016-view-online. ashx?la=en. [accessed 20/02/2017]

PDSA, 2017. Animal wellbeing report. https://www. pdsa.org.uk/~/media/pdsa/files/pdfs/veterinary/ paw-reports/pdsa-paw-report-2016-view-online. ashx?la=en. [accessed 28/09/2017]

Pet Food Manufacturer's Association. Rabbit size-o-meter. www.pfma.org.uk/_assets/docs/pet-size-o-meter/pet-size-o-meter-rabbit.pdf [accessed 06/01/17]

Podberscek, A.L., Blackshaw, J.K. and Beattie, A.W., 1991. The behaviour of group penned and individually caged laboratory rabbits. *Applied Animal Behaviour Science*, *28*(4), pp. 353–363.

Prebble, J.L. and Meredith, A.L., 2014. Food and water intake and selective feeding in rabbits on four feeding regimes. *Journal of Animal Physiology and Animal Nutrition*, *98*(5), pp. 991–1000.

Prebble, J.L., Langford, F.M., Shaw, D.J. and Meredith, A.L., 2015. The effect of four different feeding regimes on rabbit behaviour. *Applied Animal Behaviour Science*, *169*, pp. 86–92.

Rabbit Welfare Association & Fund, 2014. Understanding myxomatosis. www.rabbitwelfare.

co.uk/resources/content/info-sheets/understanding_myxo_feb06.htm [accessed 01/01/17]

Rabbit Welfare Association & Fund, 2014. Rabbits and children. https://rabbitwelfare.co.uk/about-the-rwaf/policy-statements/#children [accessed 04/10/17]

Rabbit Welfare Association & Fund, 2016. Rabbit Viral Haemorrhagic Disease update. www.rabbitwelfare.co.uk/campaign_updates/ROSpring16-CampaignUpdate.pdf [accessed 01/01/17]

Revelle, W. and Scherer, K., 2009. Personality and emotion. In D. Sander and K. Scherer (eds.), *Oxford companion to emotion and the affective sciences*, pp. 304–306. Oxford University Press.

Rodan, I. and Heath, S., 2015. *Feline behavioral health and welfare*. Elsevier Health Sciences.

Rödel, H.G. and Monclús, R., 2011. Long-term consequences of early development on personality traits: a study in European rabbits. *Behavioral Ecology*, 22(5), pp. 1123–1130.

Schalken, A.P.M., 1976. Three types of pheromones in the domestic rabbit, Oryctolagus cuniculus (L). *Chemical Senses*, 2(2), pp. 139–155.

Schepers, F., Koene, P. and Beerda, B., 2009. Welfare assessment in pet rabbits. *Animal Welfare*, 18(4), pp. 477–485.

Seaman, S.C., Waran, N.K., Mason, G. and D'Eath, R.B., 2008. Animal economics: assessing the motivation of female laboratory rabbits to reach a platform, social contact and food. *Animal Behaviour*, 75(1), pp. 31–42.

Sechi, S., Di Cerbo, A., Canello, S., Guidetti, G., Chiavolelli, F., Fiore, F. and Cocco, R., 2017. Effects in dogs with behavioural disorders of a commercial nutraceutical diet on stress and neuroendocrine parameters. *Veterinary Record*, 180(1), p. 18.

Southern, H.N., 1940. The ecology and population dynamics of the wild rabbit (Oryctolagus cuniculus). *Annals of Applied Biology*, 27(4), pp. 509–526.

Sneddon, I.A., 1991. Latrine use by the European rabbit (Oryctolagus cuniculus). *Journal of Mammalogy*, 72(4), pp. 769–775.

Speight, C., 2012. Urolithiasis – the latest thinking. *RWAF*. www.rabbitwelfare.co.uk/resources/content/info-sheets/RWAF%20Urolithiasis-calcium%20problems.pdf [accessed 02/01/17].

Tschudin, A., Clauss, M., Codron, D. and Hatt, J.M., 2011. Preference of rabbits for drinking from open dishes versus nipple drinkers. *Veterinary Record*, 168(7), p. 190.

van Praag, E., 2014. Vitamin D deficiency in rabbits. *MediRabbit*. www.medirabbit.com/Safe_medication/Vitamins/VitD_results_en.pdf [accessed 05/03/17]

Welch, T., 2015. Motivations for and thoughts toward rabbit ownership and factors contributing to companion-rabbit owners' knowledge (Master's thesis). University of Guelph, Canada.

Whary, M., Peper, R., Borkowski, G., Lawrence, W. and Ferguson, F., 1993. The effects of group housing on the research use of the laboratory rabbit. *Laboratory Animals*, 27(4), pp. 330–341.

Whitehead, M.L., 2015. '80% of entire female rabbits get uterine adenocarcinoma'. A case report of a veterinary factoid. In *Proceedings of the British Veterinary Zoological Society Meeting*, p. 37.

Whitehead, M.L., 2017. Neutering of pet rabbits. *Veterinary Record*, 180(8), pp. 204–205.

Index

rarely urinate anywhere other than their litter
trays 46
some normal behaviours are challenging to
manage 45–6
will chew and dig 46–7
will deposit faecal pellets to mark their territory
46
husbandry 11, 12
hypnotising 92, 171

interactions *see* rabbit-environment; rabbit-human;
rabbit-other species; rabbit-rabbit

jackpot rewards 140, 141, 155

latrines 119
learning
associative 134, 224
classical conditioning 134, 135, 224
operant 134–6, 224
non-associative 134, 224
punishment slows learning 139
see also training
lifting a rabbit
advice sheet 221–3
in a carry cage 221–2
training 149–50, 222
using hands 150–2
using a towel 152–3, 223
lips 110
litter training advice sheet 215–16
litter trays 43, 46, 47, 61, 119

motivations 137, 162
motivator 6, 135–6, 144, 179, 224

neutering
and care system 108
chasing behaviour 116
in females 64, 125, 164
and interactions with other species 95, 96
in males 73, 118, 119, 125–6
and mounting 81
and obesity 61
and pairings 122
preventing reproduction 122–5
and urine spraying 181
non-associative learning 134, 224
nutrition
changing diet 132–3

advice sheet 199–200
compressed 'pellets' 30
concentrate feedstuffs 27–9
monocomponent 29–31
cubes 30–1
effect of diet on behaviour 39–40
extruded 'nuggets' 30
extrusion process 29
forage feedstuffs 25
hay 25–6
straw 27
indoor advice sheet 206
outdoor advice sheet 203
overview 24–5
processing methods 29
puzzle feeding 37–8
sticks 31
training for food reward 38
treats 31

obesity 60–1
operant conditioning 134–6, 224
owners
changing/moderating expectations 2–3, 20
interaction with rabbits 44
reducing the need to pick up a rabbit 131–2
involvement in diagnosis/treatment 15
keeping a diary of behaviours 15
misconceptions around rabbit nutrition 31–7
ownership styles 7–8
perception of problem behaviours 12
taking photos/videos 16–17

pain
acute 13
case study 14
caused by environment 69
chronic 14–15
detecting 13
managing 56
parenting styles 7–8
Pavlov's dogs 134
PAW *see* PDSA Animal Wellbeing
paw contact 153
PDSA Animal Wellbeing (PAW) report (2016)
age of owners 7
behavioural problems 3, 9
changing behaviour 170, 188
environment 43
interaction with humans 44

sexual behaviour
  case study 175–6
  directed at people/objects
    females 172
    males 171–2
  mounting 175
sound
  growl 116–17
  grunt 116
  scream 117
  teeth grinding 116
  thump 117
stereotypes 188–90
stimulus 224
  behavioural 30–1
  care system 108
  desensitising/counter-conditioning 166
  differing reactions 103–4, 106–9
  environmental 68, 69, 120, 130–1, 171, 187
  fearful 18, 56, 112, 139, 142, 167
  food-related 38, 61, 168–9, 184, 190
  grooming 67, 188
  internal/external 161
  learning/training 87, 134–5, 142, 150, 182
  neutral 135, 144
  social 87, 95, 96
  unconditioned 135, 144
  unpleasant/aversive 152, 159, 179
stress 7
  clinical examination 17–19
  effect of punishment 138–9
  escalation ladder 84

teeth 35–6
  dental disease 58–9
  grinding 116

territorial aggression 125
  acquire two new rabbits at the same time 125–6
  bond current rabbit to another at home 128
  bond current rabbit to another at a rescue centre 127
time and energy budgets 13
toys 8, 35, 43, 61, 63, 86–7, 170, 202, 205, 217
training 192
  techniques 141–2
    clicker 144–8
    constructional approach 142–4
    positive reinforcement 139–41
  see also learning
trancing 87, 92–3, 171
transportation 17–18, 149
tricks 153–4

unwanted behaviours 5, 6
  welfare of animal 22
urinating 16, 46, 62, 119, 161, 178–9
  case study 180
  inappropriate 179
    fearful 179
    frustrated 179–80
    painful 179
    seeking social interaction 180
    territorial 179
    unaware 179
  spraying 180–1

welfare
  assessing 22
  Five Domains Framework 23–4
    situation-related factors 65–109
    survival-related factors 24–65
  Five Freedoms model 22–3